Renewable Energy and the Environment

Renewable Energy and the Environment

Edited by **George Thomson**

☐SYRAWOOD
PUBLISHING HOUSE

New York

Published by Syrawood Publishing House,
750 Third Avenue, 9th Floor,
New York, NY 10017, USA
www.syrawoodpublishinghouse.com

Renewable Energy and the Environment
Edited by George Thomson

© 2016 Syrawood Publishing House

International Standard Book Number: 978-1-68286-064-9 (Hardback)

Printed in the United States of America.

Contents

Preface

There is a worldwide effort by various organizations and governments of different countries to enhance the production of energy and power from diverse renewable sources. The main motive behind this shift is to reduce the harmful effects of conventional fuels on environment as well as reduce the burden on already depleting reserves of traditional fuel sources. This book attempts to elucidate the major topics of the field such as power production and supply through renewable sources, future energy resources, sustainable energy policies, trends in energy consumption from different parts of the globe, etc. The extensive content of the book would help students and researchers to delve further in this field of study.

This book has been the outcome of endless efforts put in by authors and researchers on various issues and topics within the field. The book is a comprehensive collection of significant researches that are addressed in a variety of chapters. It will surely enhance the knowledge of the field among readers across the globe.

It gives us an immense pleasure to thank our researchers and authors for their efforts to submit their piece of writing before the deadlines. Finally in the end, I would like to thank my family and colleagues who have been a great source of inspiration and support.

Editor

Solar Trigeneration: a Transitory Simulation of HVAC Systems Using Different Typologies of Hybrid Panels

Alejandro del Amo Sancho
Mechanical Department
Universidad de Zaragoza, Zaragoza, Spain
e-mail: adelamo@unizar.es

ABSTRACT

The high energy demand on buildings requires efficient installations and the integration of renewable energy to achieve the goal of reducing energy consumption using traditional energy sources. Usually, solar energy generation and heating loads have different profiles along a day and their maximums take place at different moments. In addition, in months in which solar production is higher, the heating demands are the minimum (hot water is consumed only domestically in summer). Cooling machines (absorption and adsorption) allow using thermal energy to chill a fluid. This heat flow rate could be recovered from solar collectors or any other heat source. The aim of this study is to integrate different typologies of solar hybrid (photovoltaic and thermal) collectors with cooling machines getting solar trigeneration and concluding the optimal combination for building applications. The heat recovered from the photovoltaic module is used to provide energy to these cooling machines getting a double effect: to get a better efficiency on PV modules and to cool the building. In this document the authors analyse these installations, their operating conditions, dimensions and parameters, in order to get the optimal installation in three different European cities. This work suggests that in a family house in Madrid, the optimal combination is to use CPVT with azimuthally tracking and absorption machine. In this case, the solar trigeneration system using 55 m^2 of collector area saves the cooling loads and 79% of the heating load in the house round the year.

KEYWORDS

Hybrid collectors, Solar cooling, Solar trigeneration, CPVT, PVT, Performance evaluation

INTRODUCTION

Energy savings on building applications and renewable energy technologies are two very close concepts. European regulations, like 2002/91/CE [1], are promoting energy savings in buildings through energy certificates which has been its refunded with 2010/31/UE [2] defining "Nearly Zero Energy Buildings". With the aim to reduce the energy use in buildings, it is necessary to use high efficiency installations. Several installation typologies with high efficiency are recently built, but some new installations (like geothermal, micro trigeneration, solar trigeneration, etc.) have been researched and developed, to increase the efficiency of the systems. In this work, the authors propose a comparison between some high efficiency installations.

In summer, when irradiation is high, heating demands are low, and in winter the opposite happens. Absorption and adsorption machines allow the use of the heat flow to cover cooling loads. Combining these technologies with solar thermal collectors we obtain the concept of solar cooling. In the Mediterranean climates, solar installations have many problems under summer working conditions due to the high temperatures reached in collectors. These problems may be solved using this heat flow with an absorption (or adsorption) machine. On the other hand, solar hybrid panels produce simultaneously electricity and heat which can be combined with a cooling machine, getting solar trigeneration. This hybrid system could provide the building heating and cooling loads and electricity needed in a house.

A relevant difference between an adsorption and an absorption machine is in the hot side inlet temperature coming from the collectors. Usually, absorption machines require higher temperature to work properly. Moreover, absorption machines usually have a better coefficient of performance (COP) than adsorption ones. Both machines can operate at partial loads, so a thorough study under transient conditions must be done to conclude which machine and working conditions are the more suitable depending on each climate.

Moreover, different collector typologies can be used to provide heat flow to these cooling machines. Depending on the temperature required by each machine and the partial working conditions, flat plate or concentrating hybrid collectors may be the more appropriate choice.

This study combines hybrid panels with solar cooling to get solar trigeneration system. The aim of this work is to evaluate this installation in different locations in Europe. This installation has been evaluated comparing different typologies of hybrid panels (flat plate and parabolic collectors) and different cooling machines. The transitory simulation with TRNSYS allows one to analyse the working conditions and to optimize all components in the installation (the storage tanks, solar surface, and all components in the installation).

Solar cooling has been well documented by some authors like Desideri et al. [3]. In the present document the integration of different hybrid collectors will provide more results about these installations. With this study, an optimal size of each component is found for a single-family house placed at different locations in Europe. The comparison among different typologies of panels will conclude the optimal combination for each location. The electrical efficiency in the photovoltaic cell is improved due to the cooling effect. Benefits will be evaluated, as well as working conditions using constant flow pumps and variable flow pumps.

Since the beginning of the integration of solar thermal technology on buildings in Mediterranean climates, some troubles have occurred. One of the most important is the stagnation temperature on the panels, which represents the maximum temperature reached with zero flow of the coolant. This situation occurs in the summer, when high irradiation and low demand take place at the same time. For this reason, in these climates it is interesting to use collectors with high convection losses at high temperatures. This means that a collector with high thermal loss coefficients (a_1 and a_2) will protect itself from these high temperatures in the summer.

Hybrid collectors which produce thermal energy and electricity at the same time, have been researched and documented by several authors like Chow [4], Ibrahim et al. [5] and Assoa et al. [6]. Photovoltaic modules produce around 6 - 15% of electricity (depending on the technology used), 5% is reflected, and the rest is lost as thermal energy. This thermal energy can be captured with a heat recovery system to be used in thermal applications. At the same time, the cooling effect in the solar module increases the collector efficiency (electrical and thermal). The opposite effect occurs when the application works at high temperatures, which will cause a negative effect on photovoltaic production. For cooling machines (absorption or adsorption machines), a high input temperature is required. For this reason, the hybrid collector typology will depend on the cooling machine inlet temperature.

Hybrid panels have been researched since the 80's, but are still in a state of constant development. From all existing typologies, we have selected the two most representative collectors: flat plate hybrid collector (PVT) and concentrating hybrid collectors (CPVT). Although hybrid collectors can be cooled with air or water, we have used water because it has better cooling effect in the heat recovery.

The flat plate collectors (PVT) have also several typologies as reviewed in Charalambous et al. [7], Dupeyrat et al. [8] and Dubey and Tiwari [9]. The panel considered in this work is a conventional photovoltaic module in which a copper heat recovery (sheet and tube) is installed on the back surface and an insulating layer is placed between this heat exchanger and the ambient. The module used to manufacture the PVT panel evaluated is a conventional model, with 180 W, and a temperature power coefficient of 0.45%/K. Thermal characteristics have been developed from the following equation already developed by Duffie and Beckmann [10]:

$$\eta = \frac{Q_u}{A*I_T} = F_R(\tau\alpha)_n - F_RU_L\frac{(T_i-T_a)}{I_T} - F_RU_{L/T}\frac{(T_i-T_a)^2}{I_T} \qquad (1)$$

On this equation, the main parameters F_R (heat removal efficiency coefficient) and U_L (heat loss coefficient) are 0.79 and 7.23 W/m²K respectively, which have been theoretically calculated. Usually, technical data sheets of thermal collectors contain two coefficients (a_1 and a_2, calculated from F_R and U_L) which represent how efficiency varies when the collector temperature changes. From, the PVT collector properties, calculating F_R and U_L parameters, we obtain the optical and thermal parameters with which thermal generation will be calculated. These parameters are: $\eta_0 = 0.62$, $a_1 = 5.73$ W/m²K, $a_2 = 0.00374$ W/m²K² and its correspondent efficiency curves are represented in Figure 1.

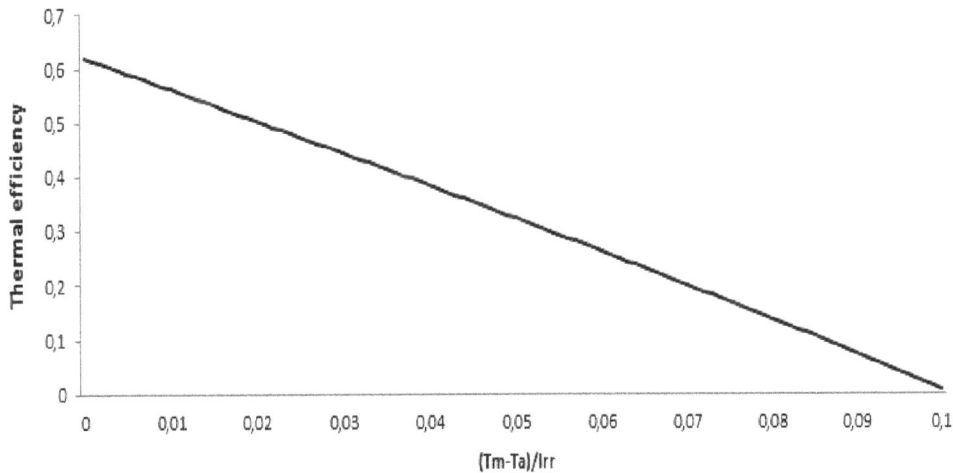

Figure 1. Thermal efficiency of a PVT (photovoltaic and thermal) panel

T_m is average temperature of the collector, T_a is ambient temperature and I_{rr} is irradiation. Concentrating hybrid collectors (CPVT) will obtain higher temperatures than flat plates. Previous studies like Mittelmann et al. [11] analyse its efficiency. On this work, a commercial collector has been evaluated [12]. This hybrid collector has a surface of 6 m², concentrating ratio of 10 times, the photovoltaic cell has a short circuit current (I_{sc}) of 13 A, an open circuit voltage (V_{oc}) of 51 V, maximum power point current (I_{mpp}) of 12.5 A, a maximum power point voltage (V_{mpp}) of 40 V and a power temperature coefficient (γ) of

0.4%/°C. CPVT collectors will not have this problem with the stagnation temperature due to the one axis (zenith) movement. In Figure 2 the CPVT studied is shown.

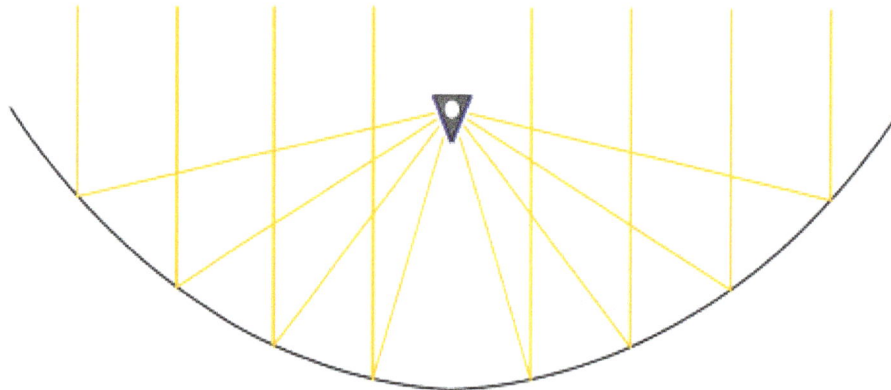

Figure 2. CPVT collector

SOLAR COOLING USING HYBRID COLLECTORS

Several installations combine thermal sources with absorption and adsorption machines like [13-15], and authors like Mittelmann et al. [11] use solar energy to activate these cooling machines. As Garcia concludes in [16] or Bermejo et al. study in Sevilla [17], solar cooling has an interesting application in Spain. The concept termed Solar Trigeneration can be achieved in different ways. Particularly, in this work it is obtained combining solar hybrid collectors with cooling machines (like absorption and adsorption). On this installation the solar cogeneration produces electricity and heat. This heat flow is used for domestic hot water or heating applications during winter and providing energy to a cooling machine in summer for air conditioning demands.

This paper aims to compare different solar hybrid collectors (PVT vs. CPVT) when they are combined with cooling machines (absorption vs. adsorption). Initially, absorption machines require higher temperatures than adsorption machines (between 85 - 95 °C and 70 - 85 °C respectively). Because of these conditions, and because the thermal efficiency in solar panel decreases when the operating temperature rises, it seems logical to use an adsorption machine. Otherwise, absorption machines have better COP than adsorption machines. Hence, adsorption machines require more energy to work and consequently more collector surface. Comparing hybrid collectors, only concentrating technology like CPVT can use the direct irradiation instead of the global irradiation which flat plate panels like PVT capture. Also, CPVT needs at least one axis tracking system, but thermal efficiency is higher than PVT panels at high temperatures. For all these concepts, in the following part we will quantify the optimum combination from the thermal efficiencies, COP and photovoltaic efficiency.

As Lecuona et al. explain in [18] the operating temperature of absorption machines is the main factor to determine the COP. To make a transient evaluation, it is necessary to consider the COP of the cooling machine at partial loads. Using the model developed using the software EES (Fig. 3) it is possible to evaluate this effect. On it, the temperatures in the generator (T_G), evaporator (T_O) and condenser (T_K) are modified.

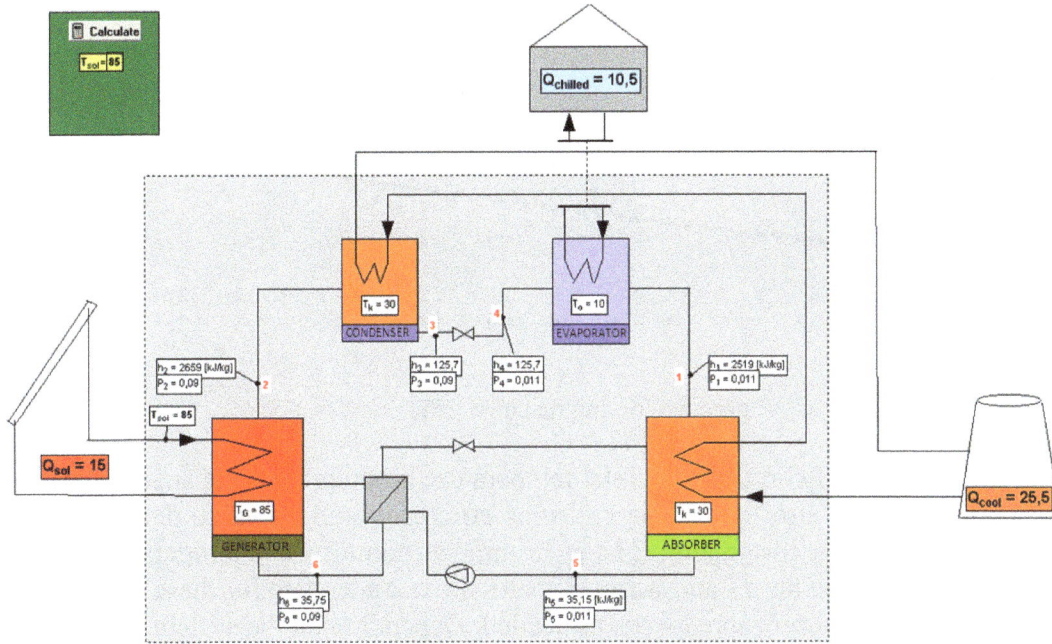

Figure 3. EES absorption machine model

As the Figures 4 - 6 show, the cooling machine performance increases when T_G is increased, when T_K is reduced and T_O is increased. Also, these figures show that it is more effective to reduce the temperature in condenser (T_K) by one degree than increasing by one degree in the generator or evaporator. The cooling machine efficiency (ε_{CM}) can be defined as the Equation 2 taken from [19]:

$$\varepsilon_{CM} = \frac{T_G - T_K}{T_G} * \frac{T_O}{T_K - T_O} \qquad (2)$$

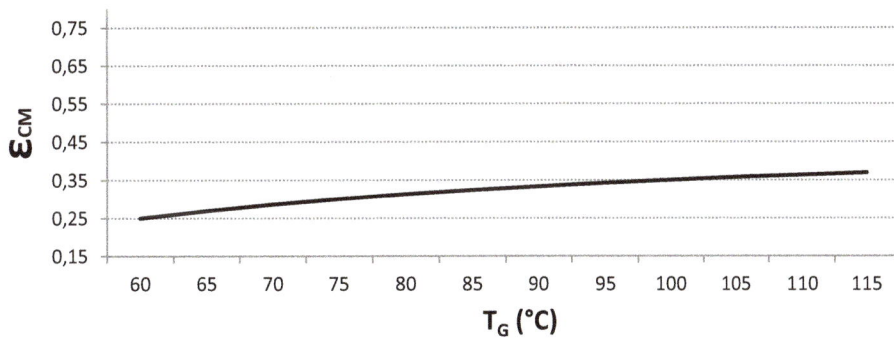

Figure 4. Cooling machine efficiency vs. T_G

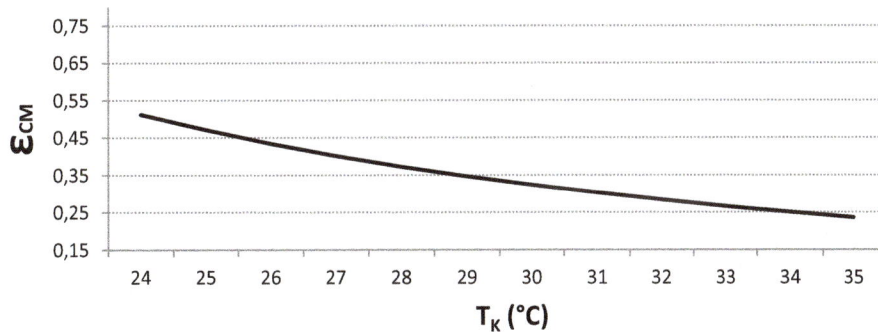

Figure 5. Cooling machine efficiency vs. T_K

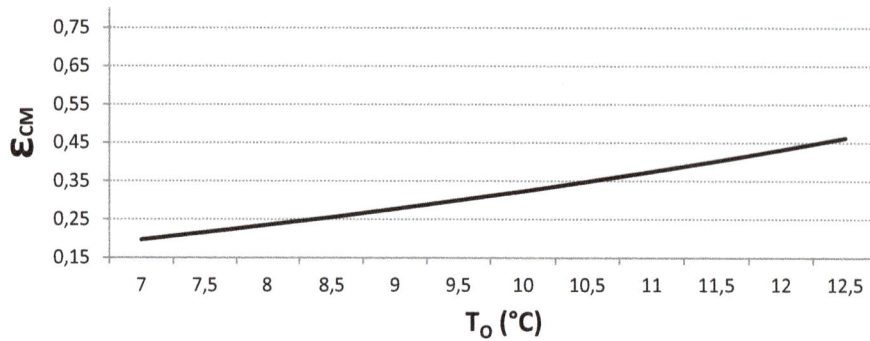

Figure 6. Cooling machine efficiency vs. T_O

T_G depends directly on the fluid inlet temperature coming from the solar collectors, and this temperature also affects the collector efficiency and its value determines the global efficiency of the installation. The environment (humidity and temperature) has an important influence on the T_K and warm climates are not adequate for these installations because the cooling tower cannot work as needed. Depending on the system used to cool the buildings, the efficiency will be different. For example, using fan-coils T_O is around 7-15 °C, or using low cooling fans around 15-17 °C, with cooling ceilings around 15-18 °C, with concrete core activation 16-19 °C, or with floor/wall cooling around 17-22 °C. Therefore, this temperature has influence on the global efficiency of the installation.

This LiBr absorption machine works with a refrigerant fluid (water) which has been simulated with Termograf software, in order to have a more accurate estimation and its diagram process, which is represented on Figure 7.

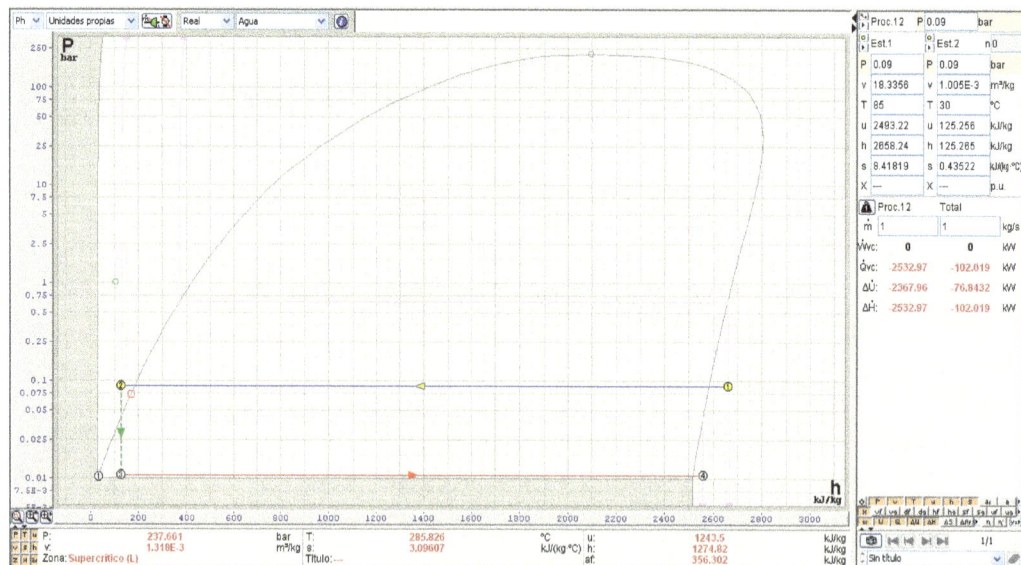

Figure 7. Refrigerant cycle calculated with Termograf

In the following parts both cooling machines, absorption and adsorption, are simulated with both collectors. To make the transient simulation, all installation has been simulated with TRNSYS, and several conclusions are taken accordingly.

Figure 8 shows the first option, where PVT panels provide a heat flow rate to an adsorption machine. Starting from the weather generator, located in Madrid, PVT panels (tilted 10°) heat the water storage tank. These tanks provide energy to a 15 kW adsorption machine which chills the cool storage tank. Both heat flow rates are dissipated through the cooling tower at the top of the figure. The first criterion is to design the hot tank with twice the capacity of the cool tank.

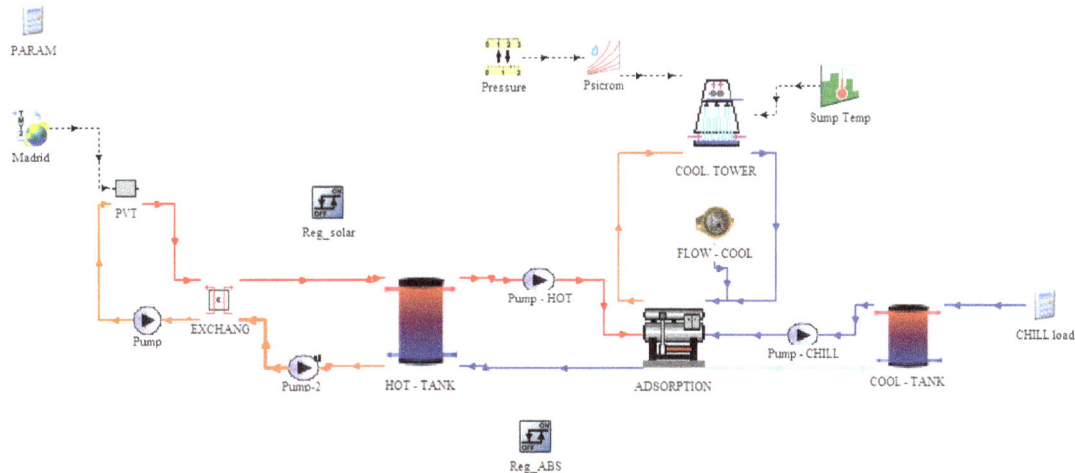

Figure 8. TRNSYS scheme – PVT with adsorption machine

To provide the energy required by this cooling machine, there are 70 m^2 of PVT collectors. Installing more collector surface will reduce the photovoltaic efficiency because the operating average temperature in the panel would be higher. This happens because the energy not provided to the cooling machine is used to increase the cell temperature. On the contrary, less collector surface means that there is not enough power in the beginning of the day which would cause intermittent working conditions. Consequently, when the solar regulator activates the pump, the flow is enough to decrease the collector temperature stopping the pump. This problem can be easily solved with a variable flow pump, with which the collector surface can be reduced by 10%. The optimal hot storage tank volume in this case is 3 m^3 and for the cool storage tank, 1.5 m^3. Higher capacities means more time to heat the hot storage tank and less volume could saturate the storage tank capacity in some periods in summer days. This saturation means that the hot tank temperature will reach its upper maximum limit (around 95 °C) and the panels will increase their temperature with a reduction in the photovoltaic generation. If the installation has thermal dissipation, this temperature won't be too high, but it will cause an additional electrical consumption. In this case the energy used to dissipate the unnecessary heat is approximately a 15% of the increased energy in the cooling effect.

Figure 9 represents during a day the thermal power generated in the collector, photovoltaic generation, heat flow rate to the adsorption machine and chilled energy supplied by it. As shown in this figure, the collectors provide energy to the storage tank from 8 AM until 4 PM, but adsorption machine is stopped by 12 PM, when cooling load is required on this day. In the beginning of the day there are some oscillations (as seen in Figure 9). This problem can be solved with a variable speed pump. Figure 10 represents the most significant efficiencies in this installation like thermal efficiency and photovoltaic efficiency in the collector and the coefficient of performance (COP). As can be seen, at the start of the cooling machine (at 12 noon), the thermal and photovoltaic efficiencies increase. In the same way, too much collector surface causes high temperatures (saturation) in the hot storage tank. These high temperatures take place because the cooling machine has a maximum power rate, and at the same time, the energy supplied by the panels to the hot storage tank is higher than the energy supplied by the storage tank to this cooling machine. Due to this energy imbalance, the hot tank heats up until 95 °C, and in this moment the solar circuit regulation stops the pumps.

Consequently, the panel reaches high temperatures and the photovoltaic cell efficiency decreases.

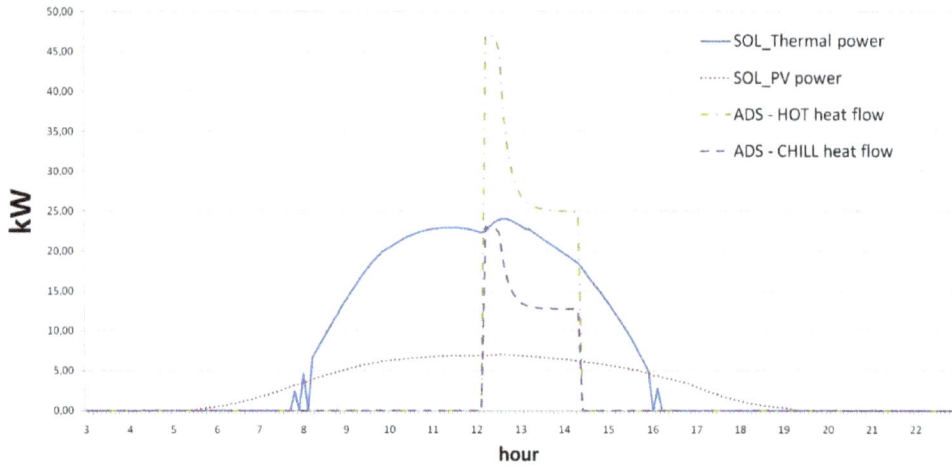

Figure 9. Daily heat flow profiles for ADS machine with PVT

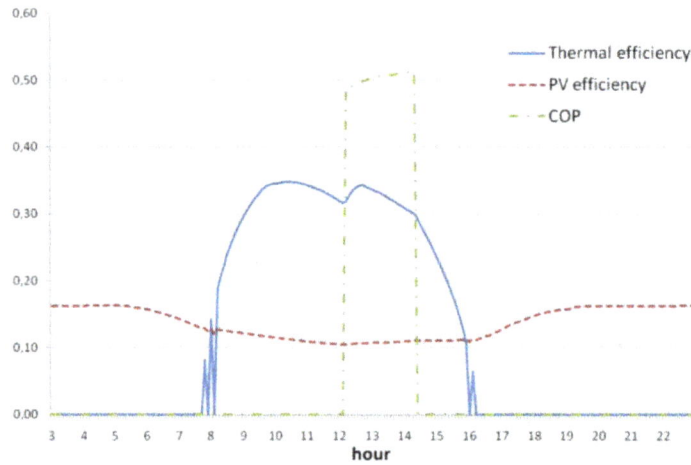

Figure 10. Installation efficiencies

The second study combines adsorption machine with CPVT. In this installation an azimuthal axis has been used, with a surface of 55 m² and a storage tank capacity of 3m³. With this collector there are no oscillations since irradiation in the collector is higher in the morning due to the incident angle. Figure 11 represents the energy daily profiles using CPVT collectors.

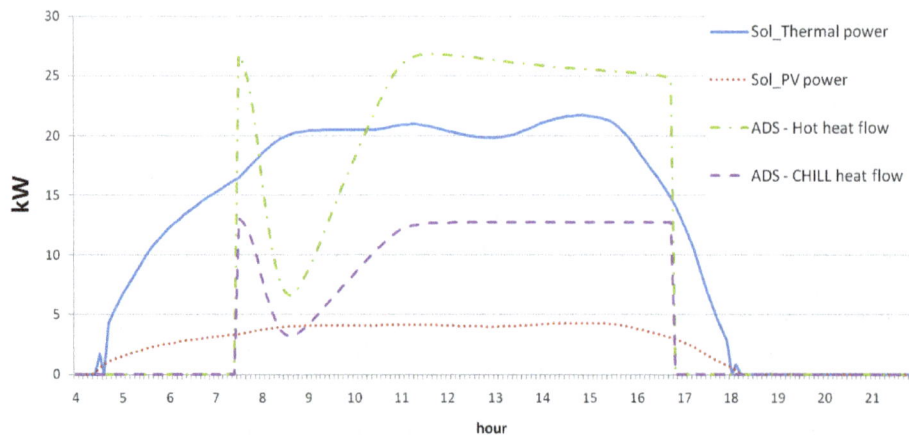

Figure 11. CPVT with adsorption machine heat flow daily profiles

An important difference between both collectors is that CPVT collectors have one axis tracker. This movement allows the capture of more irradiation at the beginning and at the end of the day. Figure 12 shows the thermal efficiency in the collectors and represent this advantage for CPVT. In Figure 13, both photovoltaic efficiencies are represented and CPVT panels usually have better values. This better efficiency takes place because photovoltaic cells used in concentrating technology have three different layers with which efficiencies around 35% can be achieved (much higher than 15 - 18% obtained in flat plate technology). Also, the thermal efficiency has better values at high temperatures (the thermal loss coefficient, usually called a_1, is lower), so it is an interesting combination with cooling machines. This photovoltaic technology is used only in concentrating collectors due to its higher price. The regulation used in the adsorption installation starts at 80 °C. This is an important effect because the adsorption will not start until the storage tank reaches this temperature, and as shown in Figure 14, using PVT panels the installation starts later. Turning some degrees to morning orientations to capture more irradiation with PVT is not enough because the slope is very small (10°). Also, Figure 15 shows the difference between outlet temperatures in both solar collectors.

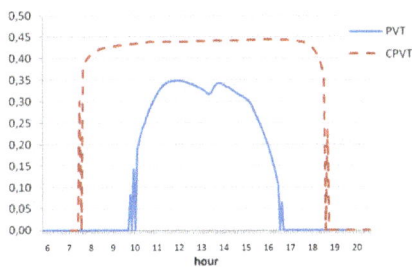

Figure 12. Thermal collectors efficiency daily profiles

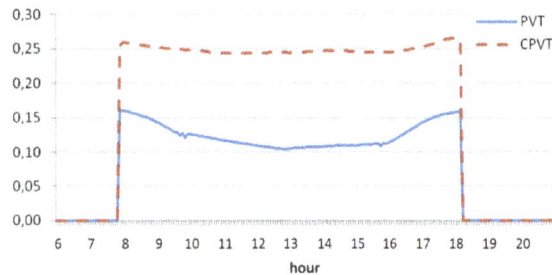

Figure 13. Photovoltaic collectors efficiency daily profiles

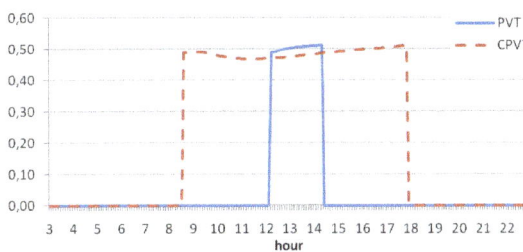

Figure 14. COP

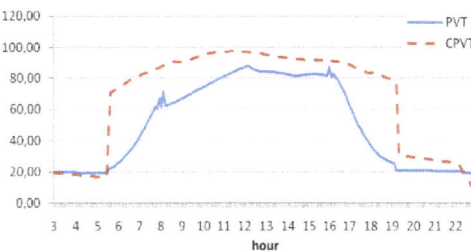

Figure 15. Solar output temperature

As explained before, absorption machines require higher temperatures than adsorption machines from collectors. This parameter (T_G) will determine the overall efficiency because it determines the solar surface needed and the cooling capacity.

The first difference between absorption and adsorption machines is the COP. High values of this coefficient can reduce notably the collector surface, providing the same energy to cooling loads. Using 70 m² of PVT collector with an absorption machine the chill power is higher than using an adsorption machine. This is due to COP difference although the thermal efficiency in the hybrid collector working at high temperatures is lower. Comparing both machines at 15 PM, absorption machine provides 18.63 kW, and an adsorption machine 12.5 kW. The conclusion is that using absorption machine with PVT panels the chill flow energy generated is higher. This means that, although a lower temperature is required for adsorption machine and consequently there is more efficiency in the collectors, the higher COP in absorption machines is more important.

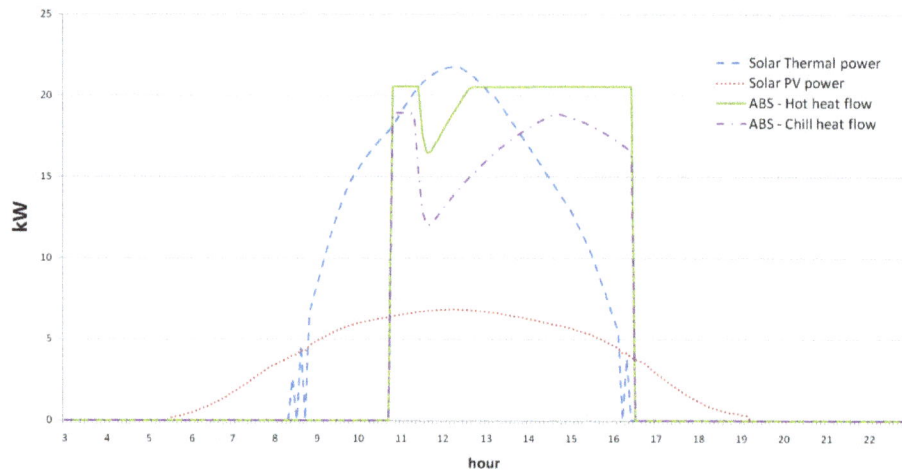

Figure 16. PVT with absorption machine heat flow daily profiles

Comparing the thermal efficiency in both collectors and machines there are three conclusions to highlight. The first one is that using PVT panels with absorption machines, thermal efficiency is lower than using adsorption machines, due to the higher temperature at which they work. On the other hand, CPVT collectors have lower differences working with both machines. Consequently, the second conclusion is that CPVT curves are more stable at temperature variations. The third difference is the required surface, 70 m² with PVT and 55 m² with CPVT. These effects are represented in the Figure 17.

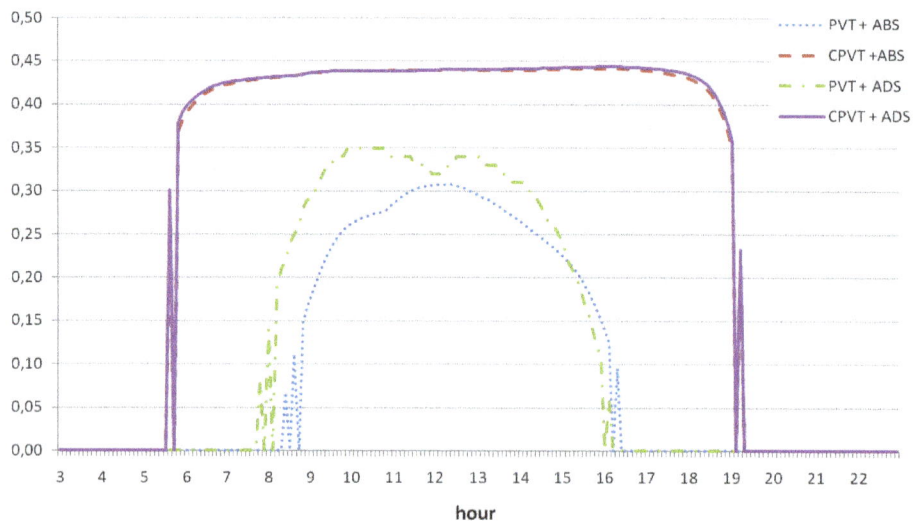

Figure17. Thermal efficiency of PVT and CPVT collectors combined with absorption and adsorption machines

From all cases studied in this paper several conclusions have drawn but, mainly there are two which must be highlighted. Using CPVT collectors, because of its required tracking system, there are higher irradiation levels in the morning and afternoon and they produce around 15% more power (in Madrid) than PVT panels. Consequently less collector surface is needed. As it has been proposed initially, some questions must be resolved in this article: which combination between collectors and machines has better global efficiency? On one side, absorption machines have higher COP, and on the other side, these machines require higher temperatures and so the efficiency in the collectors is lower. As has been explained before, better COP is more significant in the absorption machines than the loss of efficiency in the collectors. Additionally, CPVT collectors are less affected by high temperatures. So the best combination is CPVT collectors with absorption machines and the collector area required is lower.

SOLAR TRIGENERATION – YEARLY SIMULATION

Once the optimal installation has been found, the following part studies the solar trigeneration during a year. Moreover this installation will be simulated at three different locations in Europe and the aim is to determine the optimal slope for the whole year using PVT collectors.

As concluded in the previous point, the most efficient installation is the one combining CPVT panels with absorption machine. Making a transient simulation (Figure 18) during summer with 55 m² of collector, the hot storage tank temperature will not be below 75 °C. Consequently, it is not necessary to use a boiler to provide energy to the cooling machine, which would be completely inefficient. During winter, solar collectors (CPVT) contribute 79% of the heating load in a building located in Madrid, with a surface of 500 m² and with a wall thermal transmittance of 0.66 W/m²K. The same system placed in Stuttgart reduces to 18% of heating load, and a 15% in Stockholm. Hence, the proposed objective of reducing the GHG emissions has been achieved due to the high efficiency of this installation and the solar energy contribution.

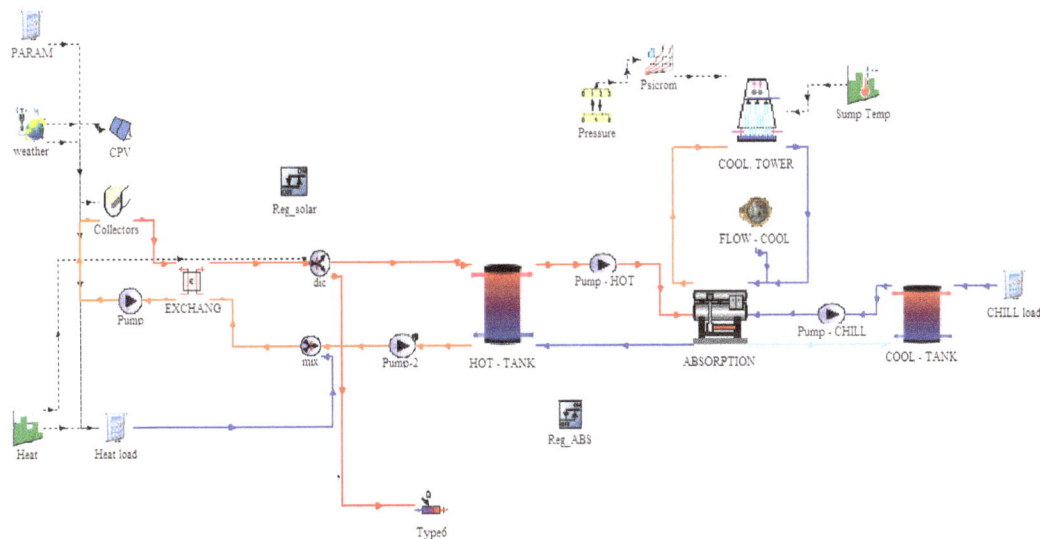

Figure 18. Heating and cooling scheme using CPVT and absorption machine

This installation covers all cooling loads with a slope of 10° and a high percentage of heating load in Madrid. To optimize the installation in winter, the tilt must be 50° in Madrid, 65° in Stuttgart and 70° in Stockholm.

Table 1 summarizes the results obtained in the simulations. First of all, energy savings in building heating loads are represented, and secondly, the optimum angle at which the collectors must be installed to get the maximum energy in the winter is shown. The installation covers all cooling loads with 55 m² collector surface, tilted by 35°.

Table 1. Heating results resume

	Madrid	Stuttgart	Stockholm
Heating savings	79%	18%	15%
Heating optimum angle	50°	65°	70°

This difference between summer and winter tilt brings a conclusion. It is more efficient to install CPVT collectors with azimuthal tracking than zenith axis. Figure 19 represents the beam irradiation incident on a surface with one axis tracking, located in Madrid. This figure shows that a collector with azimuthal axis obtains more energy in the morning and in the afternoon than a collector with a zenith axis. Because of this increment of energy, using an azimuthal axis, the cooling machines can start to operate earlier, and collector surface can be reduced to provide the same energy.

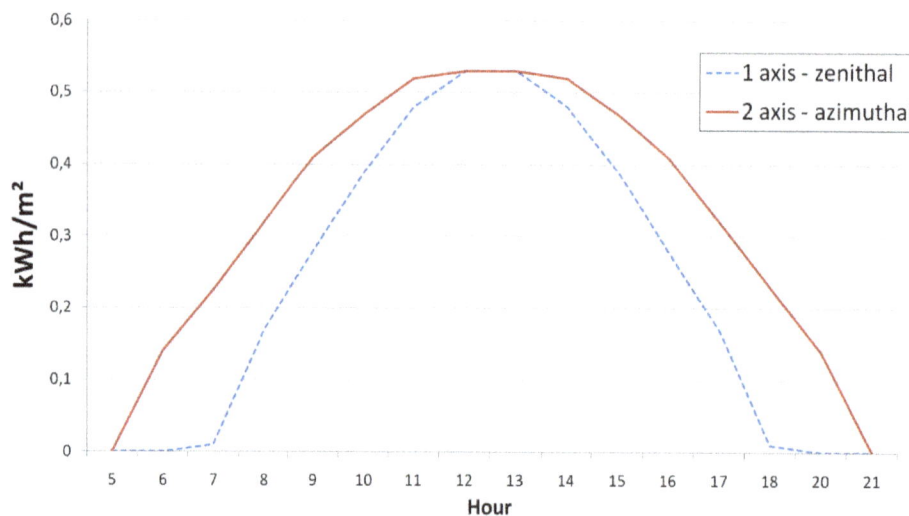

Figure 19. Irradiation daily profiles using zenithal or azimuthal axis

The energy lost during the first hours in the morning and during the last hours in the afternoon is not as important as lost collector surface in the winter because of the tilted surface. As Kostic et al. explain in [20], the angle β (collector tilted plane angle) determines notably the energy generated by hybrid collectors. Kostic optimizes with a tilt of 45° in Serbia.

CONCLUSIONS

In this paper, four solar trigeneration combinations have been compared. Two different solar hybrid (photovoltaic and thermal) panels have been evaluated using two typologies of cooling machines (absorption and adsorption). Seeing all temperatures which have influence in cooling machines, it is more effective to decrease the temperature by one degree in the condenser (T_K, from the cooling tower) than to increase one degree in the generator (T_G, provided by collectors) or one degree in the absorber (T_O, chilled heat flow). These temperatures, especially T_G will be a decisive parameter when cooling machines are combined with solar hybrid panels.

This work answers the question of which combination between hybrid collectors and cooling machines has better global performance. Comparing the thermal properties on both kind of panels proposed and the COP on each machine, the optimal combination is to use CPVT with absorption machines. The higher thermal efficiency on CPVT panels at operating temperatures (T_G) and the higher COP in absorption machines argues the conclusion. On the other hand, if we combine PVT panels with an absorption machine we have better performance than using an adsorption machine. Therefore, to get the best global efficiency, it is more relevant to have higher COPs than the thermal efficiency decrease due to the higher operating temperatures. Comparing both solar technologies, CPVT requires only 55 m^2 compared with the 70 m^2 required by PVTs.

Analysing the transient simulation we verify that during summer in Madrid, the temperature in the storage tank always exceeds 75 ºC and consequently, no auxiliary system is required. In winter, this solar surface provides enough energy to cover the 79% of the heating load. In consequence, to integrate absorption machines allows increasing solar surfaces in buildings covering much more energy demands, not only domestic hot water. As future work, authors propose to analyse the maximum solar area that is possible to use avoiding overheats in spring and autumn.

Using one axis tracker (azimuthal), the installation gets more irradiation in the morning and in consequence, the absorption machine starts around 2 hours earlier. This tracking system also allows decreasing the collector surface from 70 m² to 55 m². Using CPVT and one axis tracker in a solar trigeneration installation, the optimal tilt to maximize the yearly energy generation is 50º in Madrid, 65º in Stuttgart and 75º in Stockholm.

REFERENCES

1. Directiva 2002/91/CE del Parlamento Europeo y del consejo de 16 de diciembre de 2002 relativa a la eficiencia energética de los edificios

2. Directiva 2010/31/UE del Parlamento Europeo y del consejo de 19 de mayo de 2010 relativa a la eficiencia energética de los edificios.

3. Desideri U., Proietti S., Sdringola P., *Solar-powered cooling systems: Technical and economic analysis on industrial refrigeration and air-conditioning applications*, Applied   Energy,   86,   pp   1376-1386,   2009.

4. Chow T.T., *A review on photovoltaic/thermal hybrid solar technology*, Applied Energy, 87, pp 365-379, 2010.

5. Ibrahim A., Othmanm M.Y., Ruslan M.H., Mat S., Sopian K., *Recent advances in flat plate photovoltaic/thermal (PV/T) solar collectors*, Renewable and Sustainable Energy   Reviews,   15,   pp   352-365,   2011.

6. Assoa Y.B., Menezo C., Fraisse G., Yezou R., Brau J., *Study of a new concept of photovoltaic-thermal hybrid collector*, Solar Energy, 81, pp 1132-1143, 2007.

7. Charalambous P.G., Maidment G.G., Kalogirou S.A., Yiakoumetti K., *Photovoltaic thermal (PV/T) collectors: A review*, Applied Thermal Engineering, 27, pp 275-286, 2007.

8. Dupeyrat, P., Ménézo, C., Wirth, H., Rommel, M., *Efficient single glazed flat plate photovoltaic–thermal hybrid collector for domestic hot water system*, Sol. Energy, 2011.

9. Dubey S., Tiwari G.N., *Thermal modelling of a combined system of photovoltaic thermal (PV/T) solar water heater*, Solar Energy 2008; 82: pp 602-612.

10. Duffie & Beckman, *Solar Engineering of Thermal Processes*, John Wiley & Sons, New York, 1980.

11. Mittelman G., Kribus A., Dayan A., *Solar cooling with concentrating photovoltaic/thermal (CPVT) systems*, Energy conversion & management, 48, pp 2481-2490, 2007.
12. Bernardo L.R., Peters B., Hakansson H., Karlsson B., *Performance evaluation of low concentrating photovoltaic/thermal systems: A case study from Sweden*, Solar Energy, 85, pp 1499-1510, 2011.
13. Deng J., Wang R.Z., Han G.Y., *A review of thermally activated cooling technologies for combined cooling, heating and power systems*, Progress in Energy and Combustion   Science,   37,   pp   172-203,   2011.

14. Núñez T., Nienborg B., Tiedtke Y., *Heating and Cooling with a Small Scale Solar Driven Adsorption Chiller Combined with a Borehole System*, Task 35
15. Sözen A., Özalp M., *Solar-driven ejector-absorption cooling system*, Applied Energy, 80, pp 97-113, 2005.
16. Garcia Casals X., *Solar absorption cooling in Spain: Perspectives and outcomes from the simulation of recent installations*, Renewable Energy, 31, pp 1371-1389, 2006.
17. Bermejo P., Pino F.J., Rosa F., *Solar absorption cooling plant in Seville*, Solar Energy, 84, pp 1503-1512, 2010.
18. Lecuona A., Ventas R., Venegas M., Zacarías A., Salgado R., *Optimum hot water temperature for absorption solar cooling*, Solar Energy, 89, pp 1806-1814, 2009
19. Frío Industrial, Métodos de producción, E. Torrella, Primera edición. 2010.

20. Kostic T., Pavlovic T.M., Pavlovic Z.T., *Optimal design of orientation of PV/T collector with reflectors*, Applied Energy, 87, pp 3023-3029, 2010.

Clean Energy Certification in Brazil: A proposal

Fernando Amaral de Almeida Prado Jr.[*1], *Ana Lúcia Rodrigues da Silva*[2], *Edvaldo Marcelo Avila*[3], *Gustavo Matsuyama*[4]

[1]Faculdade de Engenharia Civil UNICAMP, Sao Paulo, Brazil
e-mail: fernando@sinerconsult.com.br
[2]Senac Campos do Jordão, Sao Paulo, Brazil
[3]Comerc Energia, Sao Paulo, Brazil
[4]Sinerconsult Consultoria Treinamento e Participações Limitada, Sao Paulo, Brasil

ABSTRACT

Given the overwhelming scientific evidence on climate change, voluntary reduction of greenhouse gas emissions is becoming increasingly important, especially as diplomatic efforts to secure the commitment of United Nations has proven ineffective. Brazil believes its energy matrix is sufficiently renewable and clean. This may be the reason why there aren't more official efforts to certify reduced emissions due to the use of clean energy sources. Nevertheless, there are financial incentives for using clean energy sources. In order to avoid the misuse of financial incentives, the electricity regulator in Brazil has developed strict supervision methodology. However, there is no data on avoided emissions. Herein we propose a methodology to certify greenhouse gas (GHG) emission reductions using best principles of governance. This methodology is based on official data, calculates the emissions avoided and the equivalent reforestation required to produce the same effect, thus making the results tangible for a less specialized audience. We also describe our practical experience with 120 consuming units that add up to over 1,500 GWh/year.

KEYWORDS

Climate change, Emissions reduction, Voluntary reduction, Reforestation, Certification.

INTRODUCTION

This article focuses on voluntary initiatives that contribute to reducing GHG emissions. Our main goal here is to develop a clean energy certification program process and encourage its use in Brazil, a country that still has no official methodology for recording and keeping track of GHG emission abatement from the use of renewable energy sources.

What drives us to this objective?

The Conference of Parties (COPs) has been unable to significantly promote diplomatic agreement, and it is unlikely that the US will adhere to international commitments unless high-emitting developing nations do the same. If emerging nations, and in particular India, China, Brazil and Indonesia adhere to the commitments, the situation could be different.

We believe that government measures and coordinated actions under the umbrella of UN agreements alone will not be enough to fight climate change. Thus, in addition to

---

[*] Corresponding author

government initiatives and related legislation, new and better market tools and voluntary self-regulation are required.

This is a hugely serious matter, as for the very first time in 3 million years the concentration of CO_2 in the atmosphere has passed the 400-ppm mark. This level means that we will have to cope with climate change for a long time, as it seems inevitable that temperatures will rise by at least 2 °C.

Thousands of references in scientific publications from all over the world confirm a worsening trend in Climate Change. One example would be the 4th IPCC Report [1]. Other examples of the many publications on this topic that describe the anthropogenic effects on climate change are those by Mann [2, 3], Dell [4], Thompson [5] and the National Research Council [6].

However, there is no unanimity about how to tackle climate change, or even if it needs to be tackled. Those who oppose the need to address climate change normally (1) discredit the existence of the climate change phenomenon, (2) recognize that climate change exists, but play down its effects and consequences, or (3) recognize the phenomenon but do not accept it is at all anthropogenic. Regardless of what they base their arguments on, they all believe there is nothing humans can do to reverse or fight the effects of climate change. Among the publications opposing or skeptical of the climate change argument are those by Lal [7], Carter [8], Idso [9], Milloy [10] and Lomborg [11].

VOLUNTARY MARKETS

Many businesses have received incentives to develop initiatives to reduce GHG emissions.

Even the US, which has yet to adhere to formal agreements and remains a paradigm of resisting international cooperation, has hundreds of initiatives to reduce emissions. Here we list some of the strategies that have been used, including by nations with no objective obligation to reduce GHG emissions:

- Regional legislation;
- Sectorial policies;
- Initiatives by industry associations, unions and NGOs;
- Business initiatives.

Each of these initiatives has its own motivations, and may or may not be facilitated by the boundary conditions of each industry.

Take for example the Carbon Disclosure Project (CDP), which in 2012 involved more than 400 companies listed in the Global 500 index. Initiatives such as this one can further best practices around the world by helping people and companies think strategically about climate change.

The 2012 CDP Report shows that 82% of the companies listed in the CDP initiative have set targets to reduce emissions. If met, these would account to 25% of the required GHG abatement to keep global warming to no more than 2 °C.

A review of the literature on voluntary markets (Desgagne, [12]; Hoffman, [13]; Delmas, [14, 15]; Kotchen, [16]; Kim [17]; Bisore, [18], Simpson [19], Hamilton [20], Fergurson [21]) indicates that information availability affects consumer concerns and consequently the demand for products.

Energy is required to produce goods and services, and consumer pressure can affect the choices people make in terms of choosing goods and services associated with renewable or clean energy [15]. This same author [14] found that a deregulated industry where competition is incipient will be affected by consumer perception and sensitivity to the issue, favoring the insertion of renewable sources.

Delmas [14] notes that the intense use of coal generation may inhibit this process because of its low cost. In the following section, we analyze the predominant sources of power in Brazil.

Kotchen [16] attempts to answer if simple, relatively low cost government programs can effectively promote voluntary initiatives, and whether such initiatives will continue to be effective if more centralized policies are required in future.

Are voluntary and mandatory initiatives complementary or substitutes? In our opinion, and likely the opinion of anyone who reads the 2013 CDP Report [22], all parties must be involved in the effort. According to Hamilton [2], the voluntary carbon market "represents a volume of over USD 100 million", and gives companies the tools they need to prepare for, and demonstrate leadership in GHG emission regulation. "This market is growing fast, perhaps doubling each year". Hamilton [20] also believes that a first step to go into a voluntary market is to measure the emissions that are the target of the effort, an opinion that is in line with this paper.

Ferguson [21], in a report for the US Environmental Agency, has a completely different view of the possibility that voluntary markets will succeed. He found that the main barriers to voluntary actions are the high cost to reduce emissions, and the complexity of the reporting disclosures.

Other authors share his pessimistic view of voluntary markets. Kim [17] reports that only projects with low marginal costs are likely to succeed, as the regulatory risk is usually considered high. This situation could be reversed if a "cap and trade" policy were used.

Finally, Hoffman [13] believes that voluntary markets could change the production chain as players opt for lower emission technologies, especially if there is some associated benefit, such as carbon credits. We would add to this the goodwill or favorable image associated with environmental marketing.

THE BRAZILIAN ELECTRIC SYSTEM

The Brazilian electric power system is characterized by a preponderance of hydroelectric power plants. These plants operate within an interconnected system, which is among the largest in the world, similar in size to the continent of Europe. This system consists of 64,000 miles of transmission lines, with planned additions of over 30,000 miles by 2021 [23]. Table 1 shows the composition of the installed generation capacity in Brazil as of January 2013 [24].

Table 1. Capacity of Brazilian power plants (January 2013)

Power plants	Capacity [MW]	Share [%]
Hydropower (HPP)	79,905.28	65.99
Small Hydro Power Plants (SHPP) (<30 MW)	4,496.32	3.71
Wind Power Plants (WPP)	1,888.38	1.56
Thermal Power Plants (TPP)	32,777.00	27.07
Nuclear Power Plants (NPP)	2,007.00	1.66
Solar Photovoltaic Power Plants (SPPP)	7.58	0.01
Total	121,081.56	100

Renewable sources of energy account for approximately 70% of capacity, while thermal power plants (including nuclear) account for the remaining 30%. However, even in hydrologically unfavorable years such as 2012, most of the power is generated by hydroelectric plants. The disproportionate role in output stems from dispatching HPPs whenever reservoirs have enough water. It is uneconomical to dispatch power plants that use costly fossil fuels (remembering that these have both direct and indirect costs in the form of GHG emissions). Table 2 shows the technology breakdown for the total output of 513.81 TWh in 2012 [25].

Table 2. Percent electricity produced by technology - Brazil, 2012

Plants	Production 2012 [%]
HPP/SHPP	85.9
WPP	0.6
TPP	10.4
NPP	3.1

Currently Brazil is the largest nation that is highly dependent on hydroelectric power. Table 3 is a ranked list published in 2010 [6] of countries producing hydroelectricity. These statistics exclude exports from Paraguay to Brazil (from its share in the power plant of Itaipu Bi-national, usually of the order of 36 TWh yearly).

This list places Brazil among the major countries with a large share of renewable electricity generation. However the acceptance of renewability as adequacy in relation to the environment is far from being accepted unanimously.

Table 3. Most significant producers of hydroelectricity

Country	Hydroelectricity production 2010 [TWh]	Share from hydroelectricity [%]
China	694	14.8
Brazil	403	80.2
Canada	376	62.0
United States	328	7.6
Russia	165	15.7
India	132	13.1
Norway	122	95.3
Japan	85	7.8
Venezuela	84	68.0
Sweden	67	42.2

The debate over the impact of hydroelectric power plants occurs in the policy arena and in academic literature. Non-governmental organizations such as Greenpeace and the World Wildlife Fund (*Movimento Gota D'água* [27], Amazon Watch [28], Sant'Ana [29]) have presented a case against electricity from hydro sources. In addition, social movements and celebrities highlight concerns regarding the dams and reservoirs and their impact on the local population and environment (flora and fauna)[†]

[†] NGOs include Brazil's MAB (Movimento dos Atingidos por Barragens) or Movement of People Affected by Dams (MAB). Celebrities include James Cameron, Arnold Schwarzenegger (AmazonWatch, 2011), and Susan Sarandon, to name a few.

Within the academic community, a sampling of those arguing the case against Hydroelectric Power Plants (HPPs) includes several examples (McCully [30], Bermann [31], Cernea [32], McCormick [33], Fearnside [34]). For these reasons, the Designated National Authority normally does not consider large hydro plants as being emission-free. However the system is almost entirely interconnected, which leads to an important question regarding the reduction of GHG emissions in regulated and non-regulated markets: how can one separate the wheat from the chaff?

Understanding the problem

The problem of how to quantify the amount of greenhouse gases emitted by interconnected power systems is a recurring one, as these systems normally combine energy produced from different sources, some of them very carbon-intense, and others that may even be considered carbon-neutral. An example of a high greenhouse gas-emitting source would be a coal burning thermal plant. A carbon-neutral source would be a small-scale wind turbine, for instance.

However, as these systems are normally operated by an independent entity, dispatch decisions are based on criteria of system efficiency and supply security, and it is impossible to unequivocally associate the dispatch of a given plant and a specific consumer.

The UN Framework on Climate Change (UNFCC) stipulates that carbon intensity shall be calculated using a set of rules based on ACM-002 methodology for calculating the average grid emission used for Clean Development Mechanisms (CDM). The methodology used to calculate the average design emission for grids under the Clean Development Mechanism (CDM) is available at http://cdm.unfccc.int/methodologies/DB/M0CSBF0F8RQG5I84XU5Y4WX0I5LHS1

In Brazil, the Designated National Authority, the entity assigned by Brazilian Law to be responsible for measures related to climate change in this country, publishes the official figures for average emissions by the Brazilian electric power grid. This calculation is updated monthly for official use in CDM projects (carbon credits) and for inventory purposes at http://www.mct.gov.br/index.php/content/view/321144.html#ancora

Although the methodology used for calculation is the same for clean development projects and inventory purposes, the values used for the former are substantially higher than those published for inventory purposes. This difference arises as calculations for carbon credit projects consider future emissions based on the expectation of new generating enterprises in the period considered for the project lifetime. In Brazil especially, there is a clear tendency towards a larger number of thermal power plants and the expectation is that emissions will increase going forward, which justifies a larger future emissions factor.

On the other hand, published emissions for inventory purposes reflect the realities of the actual generating plants dispatched. In years with less rainfall, which has been the case in 2013 and 2014, emissions are higher than they are in years with heavier rainfall. Emission factors for the Brazilian power grid are available on the Ministry of Science and Innovation Website.

The methodology described in our proposal is based on emission factors used for inventory purposes, as they reflect the actual consumption of electric power in each reference year for over 120 large electricity consumers.

The methodology to calculate avoided emissions described in this document has been used since 2011 by a portfolio of 120 electricity consumers who source part of their power from incentivized sources in the deregulated market. This portfolio includes companies in several different industries such as food, retail, electro-electronics, packaging, personal care and household cleaning, surgical supplies, construction, pulp paper, steel and metallurgy, textiles, leather, apparel, vehicles and auto parts[‡].

In the methodology used by the Brazilian Government, hydro plants of any size, biomass-fueled thermal power plants and wind farms are considered to be GHG neutral. Although it is a known fact that some hydro plants do emit greenhouse gases, the Brazilian Designated National Authority has determined that for accounting by the Brazilian electric power sector, all hydro plants shall be considered as having no GHG emissions.

Therefore, if it were possible to track the source of the power consumed one would be able to use a different emission factor for each source. In some cases, this would be higher than the average emissions of the National Interconnected System (SIN).

However, for operational strategy reasons this is impossible. Hydroplants may even be switched off to preserve water in the drier months, in which case the resulting energy deficit is offset by energy produced by thermal plants burning a range of fossil fuels.

Proposal for modeling emission reductions

By law, in Brazil small power plants using renewable resources and with low environmental impact have a financial incentive in the form of discounted energy transport rates (TUSD) or, in other words, they pay a lower fee for using the distribution systems. This incentive is determined by law 10.438/2002 which provides for the sale of electricity, [24]:

"ANEEL shall stipulate a percent reduction of not less than 50% (fifty percent) to be applied to the fee for using the power transmission and distribution systems to carry marketed electricity produced by any of the means defined in item I of this article (item I defines: the use of hydro potential greater than 1,000 kW, up to and including 30,000 kW by independent or self-generators in a small hydro plant), as well as electric power produced from wind, biomass or qualified cogeneration operations, as per ANEEL regulations and within the power ranges stipulated in cited item I."

In short, all power plants that meet the highlighted characteristics in the law above are eligible for this incentive. These include Small Hydro Power Plants, Wind Power Plants and Biomass Fueled Thermal Plants, as well as qualified cogeneration plants. All must meet the criteria of small size and necessarily be smaller than 30 MW.

Here there is a two-way match. All of the incentivized plants use renewable sources and are considered emission-neutral. Thus, by proving that the energy purchased by the consuming entity is eligible for a discount on the TUSD fee, one may infer that the entity is purchasing energy from a GHG neutral source.

The problem remains as to how to ensure that the energy used by the unit actually comes from this set of incentivized power plants.

This is guaranteed by the Electric Energy Trading Chamber, or CCEE. The CCEE is the official body responsible for supervision and control of electric energy trades between distributors, traders, free consumers and generators in Brazil. Transactions based on incentivized energy are eligible for discounted transport fees. This discount is allocated to all other electric power consumers. ANEEL assigned to CCEE the responsibility for making sure the energy traded was generated by a source eligible for the discounts provided by law. Since January 2009, the CCEE has consistently published an indicator known as the "discount matrix", with is associated with each consumer with registered contracts to purchase incentivized energy.

[‡]All of them are clients of Comerc Energia. The methodology described herein was developed by Sinerconsult Consultoria, Treinamento e Participações.

This matrix is a reliable, traceable, auditable and public measure of the energy consumed from an incentivized, and therefore GHG neutral source.

Thus, this methodology indirectly uses an official source to determine what percentage of the power consumed by a specific free consumer actually comes from a GHG neutral source.

One should remember that there is always the possibility that a given source will be unable to produce all of the energy sold. In such situations, the generator or trader must purchase energy from third parties to honor its agreements and provide the energy it sold but is unable to deliver. If this "replacement" energy comes from a non-incentivized source, the consumer loses the right to discounted transmission fees in direct proportion to the non-incentivized volume supplied. The loss of this discount is made official by the CCEE, the entity legally and formally responsible for overseeing this type of operation.

Thus the proposed methodology uses an indirect but official inspection, which identifies the proportionality of the energy with incentives and thus from sources that have zero emissions or are GHG neutral.

This methodology determines how much of the energy consumed is eligible for a transport discount, and reduces the emissions published by the government for that particular month by a proportional amount.

The outcome is supported by the reliability of two official sources, one the amount of GHG emitted by the grid, and another the exact volume of electricity consumed that was generated from renewable, GHG neutral sources. This reliability extends to the period during which the data is calculated, as both indicators are calculated for each calendar month, avoiding any distortions related to the period of calculation of these indicators.

METHODOLOGY

The method used to calculate these numbers is described below. It is based on the trading chamber (CCEE) "ME001" (energy consumed) and "EI002" (TUSD incentive discount) reports.

First the weekly consumption of energy reported in ME001 (energy consumed) reports is added up to come up with the total for the month. The amount of energy traded at a given percent discount (TUSD or wire discount) is added and divided by the total volume, to arrive at:

$$TD = \frac{\sum VE \times D}{\sum VE} \tag{1}$$

where:
- TD is the Total Discount;
- VE is the Volume of Energy;
- D is the Discount.

Consumption is then multiplied by the discount to arrive at the incentive that applies to the volume of energy:

$$MIAE = \sum WE \times TD \tag{2}$$

where:
- $MIAE$ is the Monthly Incentive Applicable Energy;
- WE is the Weekly Consumption.

The difference between monthly consumption and the amount of incentivized energy is then used to calculate the GHG emissions avoided each month, reported as tons of CO_2 equivalents. This is calculated as a specific emission factor such as tons of CO_{2eq}/MWh:

$$AE = (TMC - IAEC) \times EF \qquad (3)$$

where:
- *AE* are the Avoided Emissions;
- *TMC* is the Total Monthly Consumption;
- *IAEC* is the Incentive Applicable Energy Consumption;
- *EF* is the Emission Factor.

The procedures described herein abide by the generally accepted principles for calculating inventory, which are Relevance, Universality, Precision, Transparency and Consistency. Calculating avoided GHG emissions is a simple and reliable process if one has access to the customers reports issued by the CCEE regarding electricity consumption, specifically ME001 and EI 002.

CONCLUSION

The results obtained are substantial and represent a pioneering initiative in voluntary measures to reduce GHG emissions in Brazil.

Table 4 presents the results of 120 different companies that have been using this methodology since 2011.

Table 4. Certification results for the portfolio of COMERC clients, 2011

Segment	Ton CO_{2eq}	Equivalent reforestation (Trees)
Food	36,217	253,519.9
Retail	3,935	27,544.8
Electro electronics	793.2	5,552.2
Electromechanical	3,000.6	21,004.5
Packaging	3,274.3	22,920.4
Personal care and HH cleaning	1,373.6	9,615.0
Surgical supplies	43.6	305.2
Civil construction	4,569.4	31,985.8
Pulp & paper	2,614.2	18,299.6
Steel and metallurgy	1,738.8	12,171.8
Textile, leather and apparel	6,653.3	46,572.9
Vehicles and auto parts	1,491.4	10,439.7
Other	630.1	4,410.4
Sum	66,334.6	464,342.2

We have also calculated the equivalent amount of reforestation required to yield the same result. This should make the methodology easier for the lay public to understand, and help customers using this methodology issue sustainability reports, thus making the methodology more attractive. For this conversion, we used a reforestation project in Brazil that has received carbon credits under CDM procedures.

The procedures described herein abide by the generally accepted principles for calculating inventory, which are Relevance, Universality, Precision, Transparency and Consistency. Calculating avoided GHG emissions is a simple and reliable process if one has access to the confidential reports issued by the CCEE regarding electricity consumption, specifically ME001 and EI 002.

This methodology fills a gap in certification of avoided GHG emissions in Brazil, and may be a first step in an official certification initiative.

REFERENCES

1. Solomon, S., Qin, D., Manning, M., Chen, Z., Marquis, M., Averyt, K. B., Tignor, M. and Miller, H. L., (eds.), Contribution of Working Group I to the Fourth Assessment Report of the Intergovernmental Panel on Climate Change, Cambridge University Press, Cambridge, 2007.
2. Mann, M. E. and Jones, P. D., Global Surface Temperatures over the Past Two Millennia, *Geophysical Research Letters*, Vol. 30, No.15, pp 5-1,5-4, 2003.

3. Mann, M., Zhang, Z., Hughest, M. K., Bradley, R. S., Miller, S. K., Rutherford, S. and Ni, F., Proxy-based Reconstructions of Hemispheric and Global Surface Temperature Variations over the Past Two Millennia, *Proceedings of the National Academy of Science of the USA*, Vol. 105, No.36 pp 13252-13257, 2008.

4. Dell, M., Jones, B. F. and Olken, B. A., Climate Change and Economic Growth: Evidence from the Last Half Century, Working Paper 14132, National Bureau of Economic Research, 2008.
5. Thompson, L. G., Climate Change: The Evidence and our Options, *The Behaviour Analyst*, Vol. 33, No. 2, pp 153-170, 2010.
6. National Research Council, Climate Change: Evidence, Impacts and Choices, USA, 2012, http://nas-sites.org/americasclimatechoices/files/2012/06/19014_cvtx_R1.pdf, [Accessed: April-2013].
7. Lal, D., The New Cultural Imperialism: The Greens and Economic Development, UCLA Department of Economics, Working Paper 814, 2000.
8. Carter, R. M., Freitas, C. R., Goklany, I. M., Holland, D. and Lindzen, R. S., The Stern review: A Dual Critique, *World Economics*, Vol. 7, No. 4 pp 165-230, 2006.
9. Idso, C. and Singer, F. (orgs.), Climate Change reconsidered: 2009 Report of the Nongovernmental International Panel of Climate Change (NIPCC), The Heartland Institute, USA, 2009.
10. Milloy, S., Green Hell, Regnery Publishing Inc., USA, 2009.
11. Lomborg, B., The Skeptical Environmentalist-measuring the Real State of Earth, Cambridge University Press, 2010.
12. Desgagné, B. S. and Gozlan, E., A Theory of Environmental Risk Disclosure, *Journal of Environmental Economics and Management*, Vol. 45, pp 377-393, 2003.

13. Hoffmann, A., Climate Change Strategy: The Business Logic behind Voluntary Greenhouse Gas Reductions, Working Paper Ross School of Business, Michigan 2004.
14. Delmas, M., Russo, M. V. and Montes-Sancho, M., Deregulation and Environmental differentiation in the Electric Utility Industry, *Strategic Management Journal*, Vol. 28, pp 189 - 209, 2007.

15. Delmas, M., Montes-Sancho, M. and Shimshack, J., Information Disclosure Policies: Evidence from the Electricity Industry, *Economic Inquiry*, Vol. 48, No. 2, pp 483-498, 2010.

16. Kotchen, M., Climate Policy and Voluntary Initiatives: An evaluation of the Connecticut Clean Energy Communities Programs, Working Paper 16.117, National Bureau of Economic Research, 2010.

17. Kim, E. H. and Lyon, T. P. Strategic Environmental Disclosure: Evidence from DOE's Voluntary Green House Gas registry, *Journal of Environmental Economics and Management*, Vol. 61, pp 311-326, 2011.

18. Bisore, S. and Hecq, W., Regulated (CDM) and Voluntary Carbon Offset Schemes as Carbon Offset Markets: Competition or Complementarity? Centre Emile Bernheim, Solvay Brussels School of Economics and Management, Working Paper, 2012.

19. Simpson, P., (co-ord.), CDP Global 500 Climate Change report 2012, https://www.cdproject.net/CDPResults/CDPGlobal500-Climate-Change-Report 2012.pdf , [Accessed: May-2013].

20. Hamilton, K., Schuchard, R., Stewart, E. and Waage, S., Off Setting Emissions: A Business Brief on the Voluntary Market, Ecosystem Marketplace, 2008, http://www.bsr.org/reports/BSR_Voluntary-Carbon-Offsets-2.pdf, [Accessed: May-2013].

21. Fergurson, J., Harris, J., Hart, J. S., Ramakrishnan, K., Thompson, T. and Weber, S., Voluntary Greenhouse Gas Reduction Programs have Limited Potential, Report 08-P-0206, US Environmental Agency, Report 08-P-0206, 2008.

22. CDP, Climate Disclosure project, CDP Climate Change Report 2013, https://www.cdp.net/CDPResults/CDP-SP500-climate-report-2013.pdf, [Accessed: May-2013]

23. EPE- Empresa de Pesquisa Energetica, www.epe.gov.br, [Accessed: May-2013]

24. ANEEL - Agência Nacional de Energia Elétrica, www.aneel.gov.br, [Accessed: January-2013]

25. ONS - Operador Nacional do Sistema, www.ons.org.br, [Accessed: January-2013]

26. IEA International Energy Agency; Technology Roadmap Hydropower, 2012, www.eia.org, [Accessed: January-2013]

27. Movimento Gota d'agua, É a gota d'água, http://www.youtube.com/watch?v=WJkKoKah08A, [Accessed: February-2013]

28. Amazon Watch, James Cameron brings Arnold Schwarzenegger to Amazon to see first hand a Battle between Old and New Energy, 2011, http://amazonwatch.org/news/2011/0326, [Accessed: February-2013]

29. Sant'Ana, P. H. M., (coord.), Além das grandes Hidroelétricas: políticas para fontes renováveis de energia elétrica no Brasil- Sumário para tomadores de decisão, http://d3nehc6yl9qzo4.cloudfront.net/downloads/alem_de_grandes_hidreletricas_sumar io_para_tomadores_de_decisao.pdf, [Accessed: February-2013]

30. McCully, P., *Silenced Rivers-The Ecology and Politics of Large Dams*, London, ZED Books, 2001.

31. Bermann, C., Energia no Brasil: para quê? para quem? - Crise e alternativas para um país sustentável, São Paulo: Editora Livraria da Física / FASE v. 01. 139 p, 2002.

32. McCormick, S., The Brazilian Anti-dam movement: Knowledge, Contestation as Communicative action, *Organizational & Environment*, Vol. 19, No. 3, pp 321- 346, 2006.

33. Cernea, M. M., Social Integration and Population displacement: The Contributions of Social Sciences, *International Social Science Journal*, No. 43, pp 91-112,1995.

34.Fearnside, P., A triste história de Belo Monte III: do EIA RIMA rejeitado ao aval do Congresso, 2009, Blog do Fearnside, http://colunas.globoamazonia.com/philipfearnside/, [Accessed: February, 2013]

3

Stability Enhancement of a Power System Containing High-Penetration Intermittent Renewable Generation

Jorge Morel[*], *Shin'ya Obara, Yuta Morizane*
Department of Electrical and Electronic Engineering, Kitami Institute of Technology, Kitami, Japan
e-mail: jmorel@mail.kitami-it.ac.jp

ABSTRACT

This paper considers the transient stability enhancement of a power system containing large amounts of solar and wind generation in Japan. Following the Fukushima Daiichi nuclear disaster there has been an increasing awareness on the importance of a distributed architecture, based mainly on renewable generation, for the Japanese power system. Also, the targets of CO_2 emissions can now be approached without heavily depending on nuclear generation. Large amounts of renewable generation leads to a reduction in the total inertia of the system because renewable generators are connected to the grid by power converters, and transient stability becomes a significant issue. Simulation results show that sodium-sulfur batteries can keep the system in operation and stable after strong transient disturbances, especially for an isolated system. The results also show how the reduction of the inertia in the system can be mitigated by exploiting the kinetic energy of wind turbines.

KEYWORDS

Transient stability, Renewable energy, Wind power, Solar power, Tidal power, Storage system, Sodium-sulfur battery.

INTRODUCTION

Driven by an urgent need of CO_2 emission reduction and a strong public opinion resisting nuclear power generation following the Fukushima Daiichi nuclear disaster in 2011, Japan now has the challenge to rebuild and adapt its power system to increase its reliability and resiliency in case of natural disasters. The considered approach is based mainly on the deployment of intermittent and clean renewable energy generation, with a less centralized structure. The Japanese government has established policies to address this issue [1].

The United States of America and the European Union, in contrast, have been implementing policies to reduce CO_2 emissions by utilizing renewable energy sources even before the disaster of 2011 [2, 3].

Microgrids have evolved as a type of architecture that makes possible the generation and consumption of energy in limited and well defined areas. Their flexibility can make microgrids an active part of a larger smart grid system. Another important benefit of constructing a self-sufficient microgrid is the reduction of power to be transferred over a long distance from where centralized power plants are normally located. This reduces the losses and congestion in the transmission lines. Besides the possibility of interconnection of variable output renewable generators, there are other benefits such as increased

---
[*] Corresponding author

participation of customers in the reduction of peak demands for the entire system as well as participation in the electricity market [4].

Among the leading research teams in the field of clean energy systems is Aalborg University in Denmark, with the development of an energy system analysis tool called EnergyPLAN [5, 6]. Several studies have been performed in the design of energy systems for specific regions in Europe [7, 8] where not only electrical aspects were considered but also heat and transportation. This view is much wider than the case of smart grids alone. However, most of this work focuses on long-term operations and planning, and do not consider the dynamics of the system for normal operating condition and for the case of transient events in the electrical system.

Despite the slower development of smart energy systems in Japan, great effort has been put in by certain research groups before the nuclear disaster of 2011 [9, 10]. Recently, research activities on independent microgrids for local generation and local consumption, containing sustainable renewable energy generation such as wind, solar and tidal power have increased considerably. Operation of microgrids, aiming at the reduction of CO_2 emissions and the safety of energy supply, in cold, urban and remote areas has been studied [11-13].

In microgrids, there is a need to match instantaneous imbalances, not only for transient events in the system but also for the fast oscillations of the power outputs of renewable generators by the utilization of fast acting batteries, such as the sodium-sulfur (NaS) battery. Reference 14 analysed the utilization of NaS batteries in suppressing instantaneous or fast changing impacts in the grid, as well the possibility of independent active and reactive power control in this type of storage systems.

Storage systems have been also analysed from economic and environmental points of view [15, 16]. In [15], NaS batteries are compared to a storage system based on organic chemical hydride. However, the dynamic performance (fast charging/discharging capability) of the NaS battery should also be considered for a more complete analysis. In Reference 16, a system with no storage is analysed. Here, good CO_2 reductions are obtained despite the absence of the battery. However, for a 100% renewable supply, batteries may be necessary to shift energy between seasons or to keep frequency balance, as well as to compensate for any transient faults in the system.

For a complete analysis of a power grid, the consideration of dynamic properties of supply and demand, especially for transient events, are of vital importance. This is particularly important because of the reduction in the system inertia as wind turbines are connected to the power system through converters, which decouple the inertia of the rotating mass from the system. All the energy system designs mentioned above do not consider this aspect.

The objective of this work is to evaluate the effect of fast acting NaS batteries in the transient stability of a power system containing high penetration of renewable sources with highly variable outputs and with potential to exert strong disturbances on the system (e.g., disconnection of wind turbines due to storms) and reduced system inertia. In order to achieve this, digital simulations are performed using Matlab/Simulink.

Results of this work show that fast acting NaS batteries improve the stability level of the system, keeping the system stable after strong transient disturbances, especially for the case of isolated operation. NaS batteries can also be used for short-term balance between supply and demand during normal operating condition.

STUDY SYSTEM

In this section, the location of the study area, Kitami City, and the components of the target power system are described.

Location

Kitami City is located in a cold region, on the Hokkaido Island in the northern part of Japan, as shown in Figure 1. Kitami has an annual demand characteristic with a high heat-to-power demand ratio during winter. The temperature reaches a minimum of -20 °C in winter and a maximum of 35 °C in summer. Despite the low temperature in winter, the city gets little rainfall and snowfall. Kitami has rich natural resources such as wind, solar and tidal power that can be utilized for the generation of clean electrical energy. It is one of the richest areas in solar radiation in Japan. Also, it has open areas with good average wind speeds which can be exploited for the generation of electricity. The currents in the channels connecting the Saroma Lake and the Sea of Okhotsk offer the possibility for tidal generation [17].

Transmission network

The Japanese electricity industry consists of ten power companies that supply energy to specific and semi-independent regions. They are interconnected (except Okinawa Electric Power Company) through transmission lines with limited capacities. The Hokkaido Electric Power Company (HEPCO), shown in Figure 1, with a total installed capacity of 7,500 MW, supplies power to the Hokkaido Island, where Kitami City is located. HEPCO is connected to Tohoku Electric Power Company, located in Honshu by a High-Voltage Direct Current (HVDC) transmission system (indicated by a double line in Figure 1), with a capacity of 600 MW, approximately 8% of HEPCO's total installed capacity [18].

The Kitami City power system is connected to the local utility HEPCO which currently provides the power for the entire city. The simplified scheme of the Kitami's power system considered for simulation, including the proposed location of renewable generators and storage systems, is depicted in Figure 2. The names and rated capacities of the substations are shown in Table 1.

Selection of the location of the renewable generators was made based on available resources in the area. The location of the NaS batteries was selected to be the Rubeshibe substation where the bulk power comes from the conventional power plants of HEPCO. This selection has no direct effect in the results of the transient stability analysis presented in this paper because of the short distances involved. However, from an economical point of view, the most appropriate locations and sizes must be carefully considered.

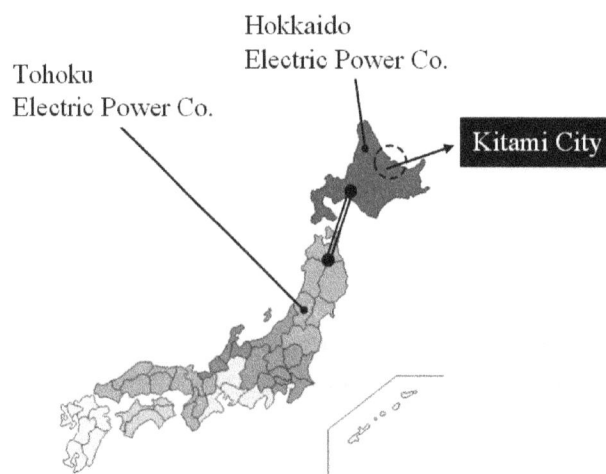

Figure 1. Kitami City location

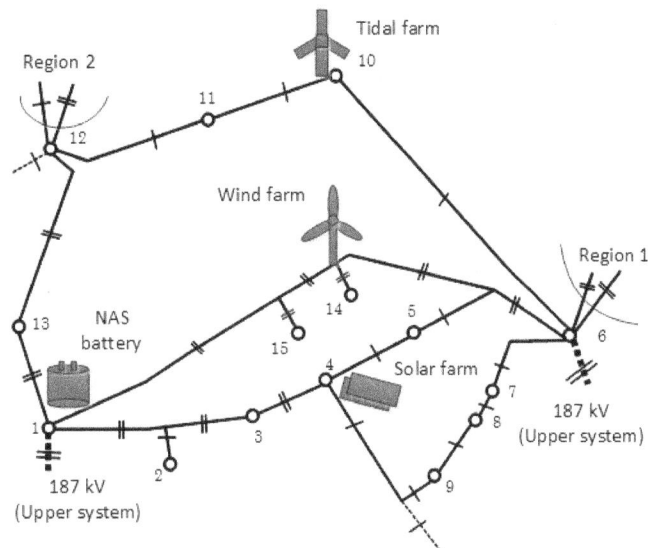

Figure 2. Power system of Kitami City

Table 1. Substation rated capacities

No.	Name	Capacity [MVA]
1	Rubeshibe	280
2	Kuneppu	6
3	Kitaminishi	20
4	Kitami	35
5	Tabata	22
6	Memanbetsu	200
7	Bihoro	20
8	Inami	10
9	Tsubetsu	12
10	Tokoro	12
11	Saroma	10
12	Engaru	18
13	Ikutahara	3
14	Kiyomi	30
15	Ainonai	12

As shown in Figure 2, there are mainly two points of connection to the local utility: The Rubeshibe and the Memanbetsu substations. Each of them is supplied by a double-circuit transmission line of 187 kV. Region 1 and Region 2 shown in this figure are two systems with no generating units. The total load for Region 1 is 124 MVA and that for Region 2 is 87 MVA. The two thin dashed lines represent the two weak connections to other systems which are not considered in this work. The small circles represent substations and single or double-circuit transmission lines are indicated by one or two transversal short lines over the lines connecting the substations. All lines are overhead with rated voltage of 66 kV and lengths of less than 40 km. Typical tower and conductor data for 66 kV transmission line is considered. Each substation is composed of 66 kV/6.6 kV step-down transformers with the rating indicated in Table 1. The reactive power of the loads is assumed to be compensated since it does not affect the present transient stability analysis.

Renewable generation

The three types of renewable generation considered in this work are wind, solar and tidal power.

Wind power. Currently, most Wind Turbines (WTs) in the market are variable-speed types: Doubly-Fed Induction Generator (DFIG) and Full-Scale Converter types, which are capable of independently controlling active and reactive power injected to the system. In this work, the wind farm is simulated using an aggregated model of DFIG-based WTs, modelled by Matlab/Simulink.

Solar power. Photovoltaic type solar farms are considered. They are connected to the transmission network via inverters which also may have the capability of independently controlling the amount of reactive power injected to the network for voltage regulation purposes. This can be exploited conveniently especially for isolated systems. Since for frequency study purposes the faster dynamics are not considered, in this work the solar farm is modelled as a first order system with a short time constant of 10 microseconds.

Tidal power. Horizontal axis tidal turbines are considered [19]. They have similar structure and working principle as the WTs [20, 21]. For short-term frequency studies the tidal turbines can be assumed to have constant power output. For long-term studies the variability can be forecasted with a high degree of accuracy. The tidal farm is simulated using an adapted version of the DFIG-based WT. These assumptions are valid due to the similarity between the wind and tidal generation systems and the time scale considered for simulation.

Horizontal wind turbines and horizontal tidal turbines have essentially the same structure: A rotor composed of blades, connected to a generator. The rotor is in charge of transforming the kinetic energy of the incoming fluid into rotating mechanical energy. The generator is in charge of transforming the mechanical energy of the rotor into electricity. The differences in both types of turbines are mainly due to the interaction between the fluids with different densities, i.e. wind and water, with the rotor. For the same rated power, a tidal turbine will have smaller rotor diameter than a wind turbine [21].

The DFIG structure is the same in both types of turbines. The generator is connected to the power grid through the stator and through the rotor. The connection of the stator is made directly or via a transformer, and that of the rotor is made by a back-to-back converter, with or without a transformer. This arrangement allows control of the electrical torque by the rotor side converter in order to adjust the rotational speed of the rotor according to wind speed variations to improve energy absorption efficiency. The detailed modelling of a tidal turbine including equations, similarities and differences with wind turbines can be found in [20, 21].

The parameters of the renewable generators are shown in Table 2.

Table 2. Aggregated renewable generator parameters

Generation	Rated capacity [MVA]	Point of connection (No.)
Wind farm	150	Kiyomi (14)
Solar farm	100	Kitami (4)
Tidal farm	1.5	Tokoro (10)

Storage systems

For long-term energy storage aiming at the seasonal and daily energy shifting, a storage system with slow dynamics but with high energy density should be utilized. For seasonal energy shifting, an organic chemical hydride type system can be employed [15].

For instantaneous and fast demand-supply imbalance compensation NaS batteries are employed due to their fast charge-discharge capability. They have been satisfactorily applied in the levelling of power outputs fluctuations of wind farms. They also can be used for load levelling and load peak shaving [22, 23].

In this work, a simplified model of the NaS battery is considered. A first order system with a small time constant of 10 microseconds, to represent the fast dynamics of this type of batteries, is considered.

CONTROL STRATEGY

For both normal and fault condition control strategies a Model Predictive Control (MPC) approach is considered for the control of the NaS batteries. Due to the uncertainty and randomness involved in the study system, MPC is used as it can handle multiple inputs and multiple outputs, in contrast to conventional proportional-integral controllers [24, 25]. Inertial control approach by WTs [26] provides primary frequency support to mitigate the reduced system inertia. The overall control strategy is shown in Figure 3.

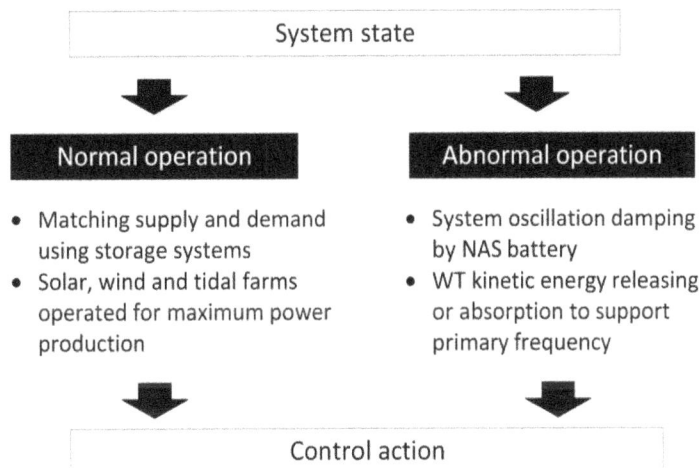

Figure 3. Overall control strategy

Normal condition

During normal condition, the WTs are controlled for maximum power absorption and maximum power output, and to limit the power output during high-speed wind conditions by applying pitch control. A similar control strategy is considered for the tidal farm. The solar farm is operated for maximum power generation. The NaS batteries are controlled to match supply and demand in the system to keep frequency within permitted ranges.

Abnormal condition

In case of abnormal or fault condition, the NaS batteries and the wind farm are operated with different strategies. Wind turbines are allowed to release or absorb, temporarily, kinetic energy by their rotating parts in order to mitigate the reduced inertia of the system and to support the primary frequency control. At the same time, battery output is entirely devoted to damp the fast oscillations of the system by absorbing or injecting active power to the network.

BUILDING OF THE SCENARIOS

Dynamic simulations are performed for three different scenarios to evaluate the effects of fast charging-discharging batteries in reducing the frequency oscillations and stabilizing the system: Target system connected to the local utility (Scenario 1), target system isolated from the local utility (Scenario 2) and primary frequency support by WTs (Scenario 3).

Scenario 1 considers the target power grid connected to that of HEPCO at two points: Rubeshibe and Memanbetsu substations. At these points, two aggregated generator models simulate the synchronous generators of the rest of the system. These generators also perform voltage and frequency control tasks.

In Scenario 2 the target power system is considered completely isolated from that of HEPCO. Only small hydropower stations in the order of 80 MVA remained connected to the system. Since there is a considerable reduction of the inertia of the system, a more severe effect is expected during transient events.

Scenario 3 shows the effect of the support provided by WTs to the network frequency during disturbances, such as in the case of a loss of an important load in the system. This scenario evaluates this WT capability assuming that the battery's stored energy level is not enough to allow any additional charge-discharge operation. A loss of 10% of the system's load is considered and the frequency profiles are analysed. The target system is assumed connected to the local utility HEPCO.

Two cases are considered for Scenarios 1 and 2: Loss of wind generation due to a storm (Case 1) and three-phase-to-ground-fault (Case 2).

In Case 1, a strong storm is assumed to affect the area where the wind farms are located. A step increase in the wind speed at $t = 20$ s is followed by the disconnection of the WTs at $t = 30$ s caused by the wind speed exceeding the cut-out wind speed level, as shown in Figure 4.

In Case 2, a three-phase-to-ground fault, with a duration of 5 cycles, is considered in the grid at $t = 35$ s, causing strong oscillations in the synchronous units of the system.

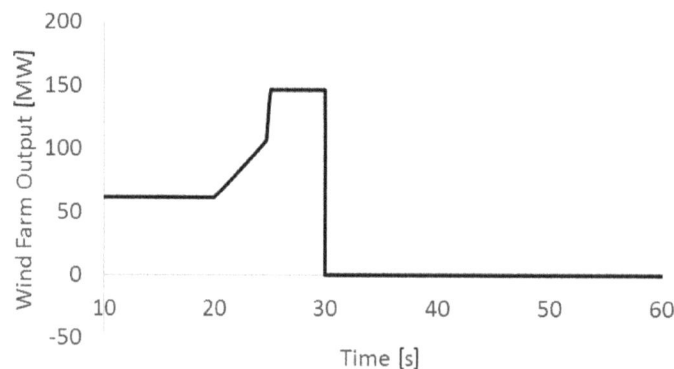

Figure 4. Wind farm power output

SIMULATION RESULTS

The overall scheme of the Matlab/Simulink model utilized for simulation is shown in Figure 5. In this figure the Wind Farm (WF), the Solar Farm (SF), the Tidal Farm (TF), the NaS battery, the two aggregated generators (HEPCO-Rubeshibe and HEPCO-Memanbetsu), together with the MPC controller are shown. The measured output (mo) is the frequency of the system (FREQ) with a reference setting of 50 Hz and the measured disturbances (md) are the WF, TF and SF outputs. The manipulated variable (mv) is the corresponding setting of the NaS battery active power. Reactive power setting of the NaS battery was considered equal to zero.

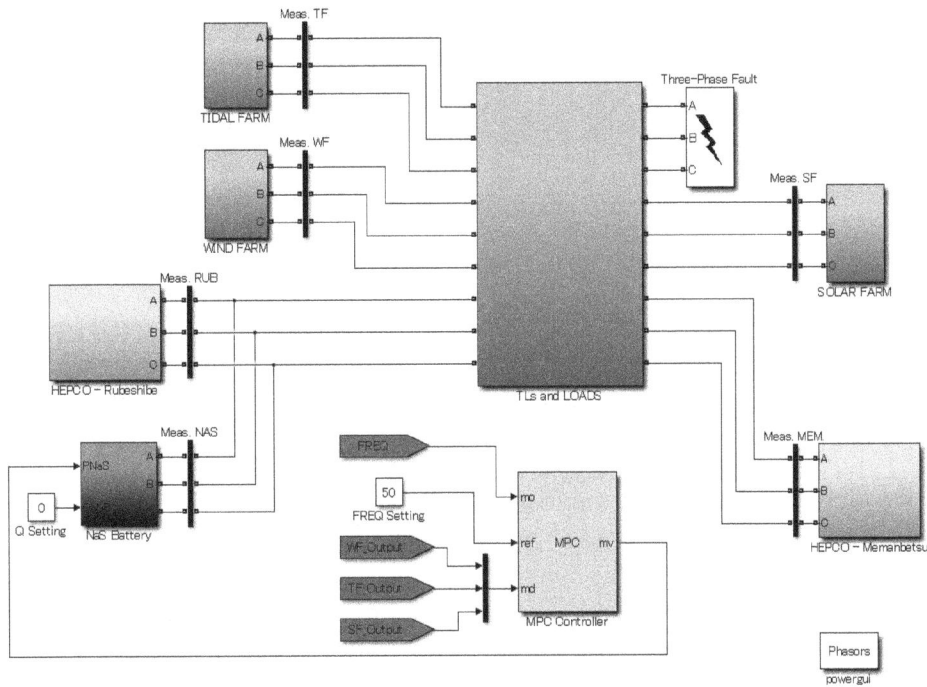

Figure 5. Matlab/Simulink model diagram

Scenario 1: Target system connected to the local utility

Case 1: Loss of wind generation due to a storm. In order to assess the stability level of the system, synchronous machine speed deviations are analysed with and without the support of the proposed control scheme. In per unit values, the speed deviations are equal to the frequency deviations. Figure 6 shows the variations in the system's frequency at Rubeshibe substation for this scenario. Without the support of the NaS batteries, the systems experiences instability. Using the fast acting batteries completely avoids the instability.

Case 2: Three-phase-to-ground fault. As shown in Figure 7, the utilization of the NaS batteries allows faster damping of these oscillations, recovering the normal frequency level, compared to the case without batteries.

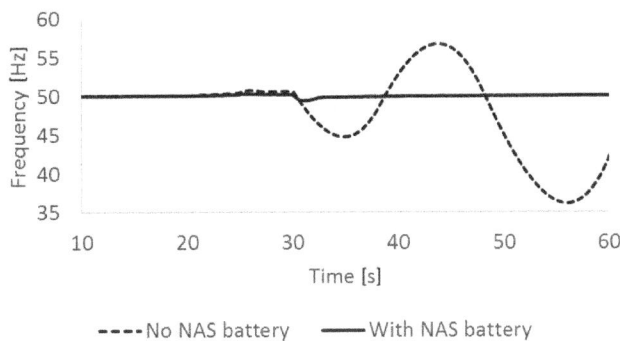

Figure 6. Frequency for loss of wind generation - Scenario 1

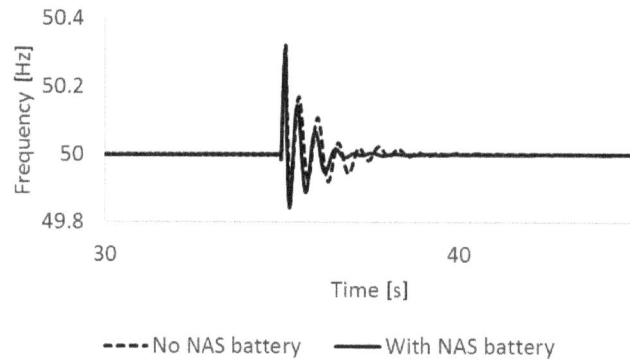

Figure 7. Frequency for fault condition - Scenario 1

Scenario 2: Target system isolated from the local utility

Case 1: Loss of wind generation due to a storm. Figure 8 shows the variations in the system frequency at Rubeshibe substation. As can be noticed, the system experiences instability if the NaS batteries are not in operation.

Case 2: Three-phase-to-ground fault. Results in Figure 9 shows that with the NaS batteries in operation the oscillations are damped completely, recovering the normal frequency level, faster than in the case when the system is not provided with this support.

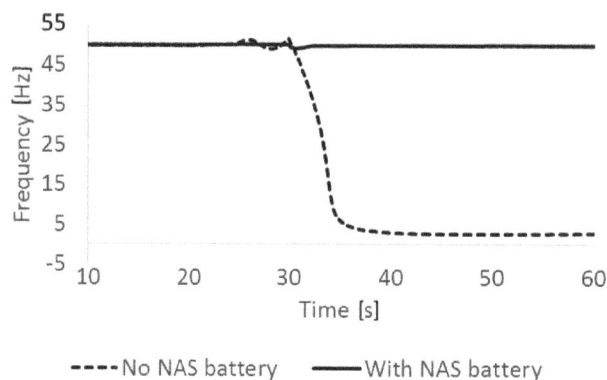

Figure 8. Frequency for loss of wind generation – Scenario 2

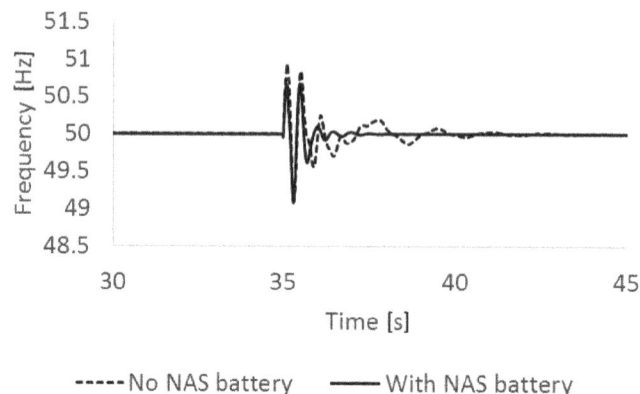

Figure 9. Frequency for fault condition – Scenario 2

Scenario 3: Primary frequency support by wind turbines

Variable speed WTs are able to release or absorb kinetic energy by their rotating parts, temporarily and almost instantaneously after a disturbance in the system, emulating the dynamics of synchronous generators and contributing to the recovery of the system's frequency. This is achieved by additional control loops to regulate WT's power output. The support of WTs is particularly useful when the level of stored energy in the NaS batteries is not adequate for proper charging-discharging operation to support system's frequency.

Figure 10 shows the primary frequency profile of the system for this scenario, with and without WT support. It can be seen that the WTs reduce the impact of a load change in the system by absorbing kinetic energy, just after the disturbance, damping the frequency peak at this point.

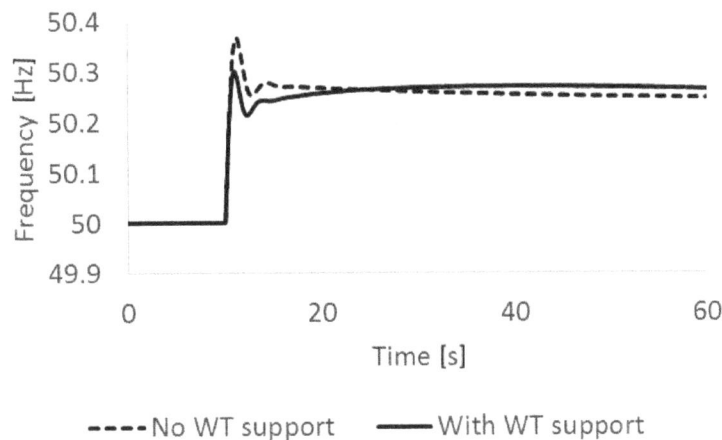

Figure 10. Frequency for loss of load - Scenario 3

CONCLUSION

This paper introduced a strategy to improve the transient stability of a power system with a high-penetration of intermittent renewable sources. The stability degradation arises mainly due to the highly variable outputs of these types of generators, during normal and abnormal operations (e.g., disconnection of wind turbines due to a storm). Reduction in the system's total inertia because of the connection of renewable generators, connected through power converters that decouple the dynamics of the generators from that of the system, also degrades system stability.

Three scenarios were considered: First, the target power system was considered connected to the local public utility, and secondly, the target power system was considered isolated. Each of these two scenarios was studied with two types of disturbances in the system: disconnection of an important wind farm due to a storm and a three-phase-to-ground fault in the power network. Finally, a third scenario was presented to demonstrate that the wind turbines can damp oscillations in the frequency by absorbing electrical energy in their rotating parts for the case of a loss of an important load in the system.

Results show that the proposed scheme, designed to charge and discharge the NaS batteries according to the oscillations in the system, and to make wind turbines participate in primary frequency support, improved the stability level by considerably reducing the synchronous machine oscillations, keeping the system operating and stable for the various scenarios presented.

In the case of more strict conditions that characterize isolated operation, voltage stability analysis must also be carefully evaluated for a more complete study of the overall stability level of the system.

REFERENCES

1. Ministry of Economy, Trade and Industry (METI), Annual Report on Energy, Outline of the FY2012 Annual Report on Energy (Energy White Paper 2013), http://www.meti.go.jp/english/report/index_whitepaper.html#energy, [Accessed: 24-February-2014]
2. U.S. Department of Energy, Mission, http://www.energy.gov/mission, [Accessed: 24-February-2014]
3. European Commission, Energy Strategy for Europe, http://ec.europa.eu/energy/index_en.htm, [Accessed: 24-February-2014]
4. SmartGrid.gov, What is a Smart Grid?, https://www.smartgrid.gov/the_smart_grid#smart_grid, [Accessed: 24-February-2014]
5. EnergyPLAN, Advanced Energy System Analysis Computer Model, Aalborg University, http://www.energyplan.eu/about/, [Accessed: 24-February-2014]
6. Lund, H., Andersen, A. N., Østergaard, P. A., Mathiesen, B. V., Connolly, D., From Electricity Smart Grids to Smart Energy Systems - A Market Operation Based Approach and Understanding, *Energy – The International Journal, Elsevier*, Vol. 42, No. 1, pp 96-102, 2012.
7. Connolly, D., Lund, H., Mathiesen, B. V., Werner, S., Möller, B., Persson, U., Boermans, T., Trier, D., Østergaard, P. A., Nielsen, S., Heat Roadmap Europe: Combining District Heating with Heat Savings to Decarbonise the EU Energy System, *Energy Policy, Elsevier*, Vol. 65, pp 475-489, 2014.

8. Connolly, D., Lund, H., Mathiesen, B. V., Leahy, M., The First Step Towards a 100% Renewable Energy-system for Ireland, *Applied Energy, Elsevier*, Vol. 88, No. 2, pp 502-507, 2011.
9. Obara, S., Development of a Dynamic Operational Scheduling Algorithm for an Independent Micro-Grid with Renewable Energy, *Journal of Thermal Science and Technology, The Japan Society of Mechanical Engineers (JSME)*, Vol. 3, No. 3, pp 474-485, 2008.
10. U. S. Department of Energy, Microgrids at Berkeley Laboratory, Nagoya 2007 Symposium on Microgrids, Overview of Micro-grid R&D in Japan, http://der.lbl.gov/sites/der.lbl.gov/files/nagoya_morozumi.pdf, [Accessed: 25-February-2014]
11. Cellura, M., Di Gangi, A., Orioli, A., Assessment of Energy and Economic Effectiveness of Photovoltaic Systems Operating in a Dense Urban Context, *Journal of Sustainable Development of Energy, Water and Environment Systems*, Issue 1, No. 2, pp 109-121, 2013.
12. Quoilin, S., Orosz, M., Rural Electrification through Decentralized Concentrating Solar Power: Technological and Socio-Economic Aspects, *Journal of Sustainable Development of Energy, Water and Environment Systems,* Issue 1, No. 2, pp 199-212, 2013.
13. Obara, S., Kawai, M., Kawae, O., Morizane, Y., Operational Planning of an Independent Microgrid Containing Tidal Power Generators, SOFCs, and Photovoltaics, *Applied Energy, Elsevier*, Vol. 102, pp 1343-1357, 2013.

14. Ohtaka, T., Iwamoto, S., A Method for Suppressing Line Overload Phenomena Using NaS Battery Systems, *Electrical Engineering in Japan, Wiley*, Vol. 151, No. 3, pp 19-31, 2005.

15. Obara, S., Morizane, Y., Morel, J., Economic Efficiency of a Renewable Energy Independent Microgrid with Energy Storage by a Sodium-Sulfur Battery or Organic Chemical Hydride, *International Journal of Hydrogen Energy, Elsevier*, Vol. 38, No. 21, pp 8888-8902, 2013.

16. Obara, S., Morel, J., Microgrid Composed of Three or More SOFC Combined Cycles without Accumulation of Electricity, *International Journal of Hydrogen Energy, Elsevier*, Vol. 39, No. 5, pp 2297-2312, 2014.

17. Japan Meteorological Agency, http://www.jma.go.jp/jma/indexe.html, [Accessed: 26-February-2014]

18. Hokkaido Electric Power Company, Main Infrastructure, http://www.hepco.co.jp/corporate/ele_power/ele_power.html, [Accessed: 26-February-2014] (In Japanese)

19. Tocardo Tidal Turbines, http://www.tocardo.com/, [Accessed: 26-February-2014]

20. Aly, H. H. H., El-Hawary M. E., Comparative Study of Stability Range of Proposed PI Controllers for Tidal Current Turbine Driving DFIG, *International Journal of Renewable and Sustainable Energy*, Vol. 2, No. 2, pp 51-62, 2013.

21. Ben Elghali, S. E., Benbouzid, M. E. H., Charpentier, J. F., Modeling and Control of a Marine Current Turbine Driven Doubly-Fed Induction Generator, *IET Renewable Power Generation*, Vol. 4, No. 1, pp 1-11, 2010.

22. Tanaka, K., Kurashima, Y., Tamakoshi, T., Recent Sodium Sulfur Battery Application in Japan, Bonneville Power Administration, http://www.bpa.gov/Energy/n/tech/energyweb/docs/Energy%20Storage/NGK-Paper.PDF, [Accessed: 26-February-2014]

23. NGK Insulators Ltd, http://www.ngk.co.jp/english/products/power/nas/, [Accessed: 26-February-2014]

24. Bemporad, A., Morarim M., Ricker N. L., Model Predictive Control Toolbox User's Guide, *The MathWorks, Inc.*, 1998.

25. Ernst, D., Glavic, M., Capitanescu, F., Wehenkel, L., Reinforcement Learning Versus Model Predictive Control: A Comparison on a Power System Problem, *IEEE Transactions on Systems, Man, and Cybernetics*, Vol. 39, No. 2, pp 517-529, 2009.

26. Mauricio, J. M., Marano, A., Gomez-Exposito A., Martinez Ramos J. L., Frequency Regulation Contribution Through Variable-Speed Wind Energy Conversion Systems, *IEEE Transactions on Power Systems*, Vol. 24, No. 1, pp 173-180, 2009.

4

Solar Distillation System Based on Multiple-Effect Diffusion Type Still

*Bin-Juine Huang[*1], Tze-Ling Chong[1], Hsien-Shun Chang[1],
Po-Hsien Wu[1], Yeong-Chuan Kao[2]*

[1]Department of Mechanical Engineering
National Taiwan University, Taipei, Taiwan
e-mail: bjhuang@seed.net.tw

[2]Department of Physics
National Taiwan University, Taipei, Taiwan

ABSTRACT

The present study intends to develop a high-performance solar-assisted desalination system (SADS) using multi-effect diffusion type still (MEDS) and the vacuum tube solar collector (VTSC). A MEDS prototype was designed and built. The measured result is very close to the estimation. The 10-effect MEDS will produce pure water at about 13.7 L/day/m^2 collector area at a solar irradiation of 600 W/m^2 and 19.7 L/day/m^2 collector area at solar irradiation 800 W/m^2. For 20-effect still, the yield rate increase is 32% compared to 10-effect still.

KEYWORDS

Solar desalination, Solar still, Multi-effect diffusion, Multi-effect diffusion solar still, Solar water purification, Solar distillation

INTRODUCTION

The multi-effect diffusion still (MEDS) is a high-efficiency heat-driven distillation device, which was published by Cooper and Appleyard in 1967 [1, 2]. Although some researchers developed MEDS for experiment [3-8], not many good application devices based on the principle of MEDS are available due to the problem of suitable manufacturing techniques. The present study intends to develop a high-performance solar-assisted desalination system (SADS) in which MEDS is adopted.

The MEDS (Figure 1) contains a number of vertical cells which are closely spaced. One face of each cell is a plain plate on which vapour condenses. A thin layer of capillary material (wick) is attached to the other side of this plate to act as a wick. Seawater is supplied to the wick at the top. The heating plate absorbs the heat released from the condensation of water vapour vaporized from the saturated wick of the upstream cell. The condensation heat is then conducted through the heating plate to heat the seawater inside the wick and produce vapour. The vapour diffuses towards and condenses at the next heating plate. The collectors under the heating plates collect the drops of pure water. Latent heat released during the condensation on the bare side of the heating plate evaporates the impure water which has saturated the wick material on the other side of the

---

[*] Corresponding author

heating plate. The process repeats until the last cell and the residual heat is finally discharged to the environment.

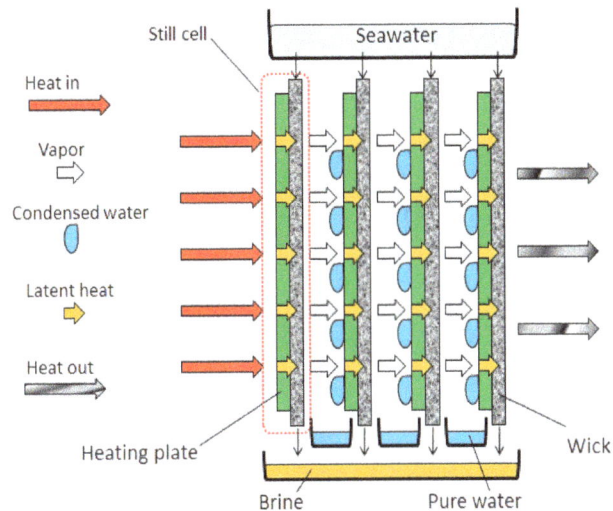

Figure 1. Multi-effect diffusion type still (MEDS)

The design principle of MEDS is perfect. But the vapour diffusion process in cells is difficult due to low convective heat and mass transfer and little vapour diffusion potential exists in each cell. The temperature gradient of each cell becomes smaller when more cells are used. There must be an optimal design on the total number of cells for the best performance. A lot of researches including innovative designs need to be done in order to make MEDS applicable.

For SADS (Figure 2), the vacuum-tube solar collector (VTSC) is used to supply heat. The solar heat absorbed by VTSC evaporates the water and produces high temperature steam as the heat source of MEDS. The vapour generated by solar heat condenses at the first heating plate of MEDS cell, releasing the latent heat in the distillation process. The hot salt brine product and hot pure water are used to preheat the input seawater. VTSC is designed using an automatic water feeding mechanism. When the VTSC is cooled down at night, by the function of check valves, it creates a negative pressure in the water storage which sucks the water into the storage spontaneously.

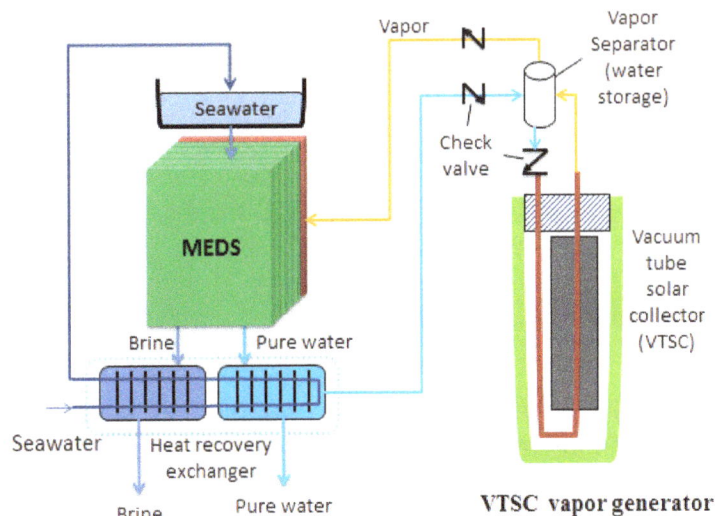

Figure 2. Solar-assisted desalination system (SADS)

DESIGN ANALYSIS OF MEDS

For commercialization purpose, the design guidelines include low-cost, high efficiency, high reliability, and simple maintenance. The concept of modular design is employed. Especially for the components with short life time such as still cell, low-cost and replaceable design is necessary. The wicks of the still cells undergo degradation easily due to pollution by seawater. The design of cells should be in modular type for easy replacement.

The solar collector used for SADS is the module with 6 tubes of VTSC (Figure 3) and the total absorber area is 0.92 m². The thermal efficiency of this VTSC module is about 0.5 at 100 °C.

Figure 3. VTSC module

Table 1 shows the estimation of steam production rate at different levels of solar irradiation.

Table 1. Steam production estimation of VTSC module

Solar irradiation, [W/m²]	400	600	800
Vapour production rate, [g/min]	4.9	7.4	9.8

The SADS is designed for seawater desalination. Since the concentration of salt in seawater is high, to avoid salt crystallization in the wick, there is a minimum seawater flow through the wick. The salt concentration of seawater is 0.036 g/ml, and the saturated salt solubility in water is 0.36. From mass balance, we can derive the ratio of the production of pure water and that of salt brine. From the mass balance of total mass flow, we obtain:

$$m_i = m_v + m_e \tag{1}$$

where m_i is the total mass flow rate at the inlet; m_v is the mass flow rate of the evaporated vapour; m_e is the mass flow rate at exit (brine). From the mass balance to the salt, we obtain

$$C_i m_i = C_e m_e \text{ or } m_i = C_e m_e / C_i \tag{2}$$

where C_i is the salt concentration at the inlet flow (0.036); C_e is the salt concentration of exit (brine) (0.36). Combining Equations 1 and 2, we obtain the ratio of pure water vapour flow to the brine flow:

$$r = \frac{m_v}{m_e} = \frac{C_e - C_i}{C_i} \tag{3}$$

The theoretical analysis of heat and mass transfer of the MEDS is similar to [3-7] with a constraint that the ratio r should be less than 9 in order to avoid salt crystallization in the wick.

A simple analysis was carried out in the present study. Within two adjoining still cells, the heat transfer from cell to cell is by three mechanisms: conduction, convection, and diffusion. Ignoring the heat leak from the edges of cell, the diffusion transfer will increase with decreasing gap between the cells. For the MEDS studied, the diffusion gap is small enough such that the diffusion process dominates the other transport processes, and the influence of conduction and convection can be ignored. That is, the heat transfer from cell to cell is dominated by water evaporation, diffusion, and vapour condensation and the major heat loss is from the overflow of salt brine. The simplified energy conservation equation for a single still cell then becomes:

$$q_i = c_p \Delta T\, m_i + h_{fg} m_v \qquad (4)$$

where q_i is the heat input, h_{fg} is the latent heat of water, c_p is the specific heat of water and ΔT is the temperature difference between input seawater and output salt brine. The mass flow rate of condensate, i.e. pure water production rate m_{vo}, is equal to the mass flow rate of evaporation:

$$\mathrm{m}_{vo} = m_v \qquad (5)$$

Combining Equations (3) to (5), we can estimate the pure water production rate for a single still cell from the following equation:

$$\mathrm{m}_{vo} = \frac{q_i}{(r+1)c_p \Delta T + h_{fg}} \qquad (6)$$

The thermal resistances of each still cell, which are inversely proportional to the heat transfer area, cause the temperature gradient from cell to cell. To simplify the analysis, we assume that the temperature decreases uniformly from the heating plate of the first cell to the last. The total pure water production rate from all still cells can then be predicted. The total pure water yield per day (6 hours) for one module of VTSC (with solar thermal energy efficiency 0.5) can also be predicted for different values of solar irradiation.

A performance index is defined in the present study for performance evaluation. The solar energy coefficient of performance (R_S) is defined as

$$R_S = \frac{\text{total vapor condensation heat required for actual pure water production}}{\text{solar energy input to VTSC}} \qquad (7)$$

For the evaluation of the distillation process alone excluding the solar energy conversion efficiency, we define the distillation coefficient of performance (R_o) as

$$R_o = \frac{\text{total vapor condensation heat required for actual pure water production}}{\text{heat input to MEDS}} \qquad (8)$$

Table 2 shows some estimated results of 10-effect MEDS, and Table 3 for 20-effect MEDS, with $r = 7$ and $T_i = 30\ ^\circ$C. The daily average solar irradiation in Taiwan is about

600 W/m^2 (12.9 MJ/m^{2}day) (in summer) and is about 800 W/m^2 (17.3 MJ/m^{2}day) in desert area. If there is no heat recovery, for 10-effect MEDS, the daily pure water production rate (for 6 hours) is around 12.6 L/day/set (13.7 L/m^2/day) in Taiwan and 18.1 L/day/set (19.7 L/m^2/day) in desert area, where the solar absorber area of one set VTSC is 0.92 m^2. This value is much higher than the stills we have ever made.

Table 2. Pure water production estimation of 10-effect MEDS

Solar Irradiation, [W/m^2]	10-eff Pure water yield, [L/day/set] without heat recovery											R_S
	1st	2nd	3rd	4th	5th	6th	7th	8th	9th	10th	Total	
400	1.02	0.91	0.82	0.75	0.69	0.64	0.61	0.58	0.57	0.55	7.13	2.02
600	1.8	1.6	1.44	1.32	1.22	1.14	1.08	1.03	1	0.98	**12.6**	2.37
800	2.57	2.3	2.07	1.89	1.75	1.63	1.55	1.48	1.43	1.41	**18.1**	2.55
	10-eff Pure water yield, [L/day/set] with heat recovery											
400	1.06	0.98	0.92	0.87	0.83	0.81	0.8	0.8	0.8	0.8	8.65	2.44
600	1.86	1.73	1.62	1.54	1.48	1.43	1.41	1.41	1.41	1.41	**15.3**	2.88
800	2.67	2.48	2.32	2.2	2.12	2.06	2.02	2.02	2.02	2.02	**21.9**	3.1

For 20-effect MEDS, the daily pure water production rate (for 6 hours is around 16.5 L/day/set (17.9 L/m^2/day) in Taiwan and 23.7 L/day/set (25.8 L/m^2/day) in desert area. For 20-effect still, the yield rate increase is 32% compared to 10-effect still. All the above results will be improved if the heat recovery device was added.

In reference [7], Tanaka proposed a MEDS coupled with heat-pipe solar collector. The daily water production of Tanaka's solar still with different number of effects was determined by theoretical analysis [7], Figure 7. For 10-effect still with solar radiation 24.4 MJ/m^{2}day, the daily water production of Tanaka's still is around 20 L/m^{2}day. Assuming that the thermal efficiency of VTSC in present study is same as the solar collector of Tanaka's still, for similar weather condition and seawater feeding rate, the present theoretical prediction of daily water production in the 10-effect MEDS is 18 L/m^{2}day, which differs by 10% compared with Tanaka's result.

Table 3. Pure water production estimation of 20-effect MEDS

Solar Irradiation	20-eff Pure water yield [L/day/set] without heat recovery											R_S
	2nd	4th	6th	8th	10th	12th	14th	16th	18th	20th	Total	
400 [W/m^2]	0.89	0.7	0.57	0.47	0.4	0.4	0.31	0.28	0.26	0.25	9.36	2.65
600 [W/m^2]	1.57	1.24	1	0.83	0.7	0.7	0.54	0.5	0.46	0.44	**16.5**	3.12
800 [W/m^2]	2.25	1.78	1.44	1.19	1.01	1.01	0.78	0.71	0.66	0.63	**23.7**	3.36
	20-eff Pure water yield [L/day/set] with heat recovery											
400 [W/m^2]	0.96	0.82	0.71	0.64	0.58	0.58	0.53	0.52	0.52	0.52	13	3.68
600 [W/m^2]	1.69	1.44	1.26	1.13	1.03	1.03	0.93	0.92	0.92	0.92	**22.9**	4.33
800 [W/m^2]	2.43	2.07	1.81	1.62	1.48	1.48	1.34	1.32	1.32	1.32	**32.9**	4.66

DESIGN OF MEDS

Figure 4 shows the design of MEDS. All the components can be assembled and disassembled quickly. Twenty still cells can be installed in a unit of MEDS.

Figure 4. Design of MEDS

The vapour collector receives the vapour from vapour generator (e.g. solar collector). The wick is made of porous material which has high permeability and capillary effect.

The seawater supply contains a water distributor. The feeding rate can be controlled using different materials or adjusting the liquid level in the supply tank. The distributor can be replaced regularly.

The design of the heat recovery device is simply a water pipe covered by capillary material. The heat exchange occurs between the liquid in the capillary (input flow) and the liquid inside the pipe (pure hot water) to recover the heat.

The dimension of the still cell plate is 30 cm x 40 cm x 0.5 mm, which is made of polycarbonate plastic. The separating distance between the adjacent still cells is 6 mm. The still cells are handmade in lab for the experiment. Fabrication by moulding can be applied in mass production.

Figure 5 shows the external appearance of a MEDS unit. The dimension of the casing is 67 cm x 28 cm x 45 cm. The pure water production capacity of one MEDS unit is to match with 3 tubes of VTSC at solar irradiation 600 W/m^2. For 1 set of VTSC, 1 to 3 units of MEDS are needed in different weather conditions.

Figure 3 shows the VTSC vapour generator used in the experiment, which is a flow-through type vacuum-tube collector. The vapour separator is added to the vacuum-tube collector as an accumulator and separator. The accumulator volume is 2.5 L.

BASIC PERFORMANCE TEST OF MEDS

The performance test of the MEDS prototype was carried out to understand the process of heating, evaporating, condensation, and flows at different parts. The vapour production rate of the VTSC has been tested at different solar irradiation levels. Table 4 shows that the measured results are very close to the estimated results. The uncertainty of the values is due to the unstable solar irradiation. The thermal performance test of VTSC vapour generator shows that the solar thermal conversion efficiency is about 0.6.

Table 4. VTSC vapour generator test

Solar irradiation, [W/m^2]	Measured, [g/h]	Estimated, [g/h]
400	180-300	294
600	360-480	444
800	480-600	588

The MEDS distillation tests are carried out using an electric steam generator. Table 5 shows some test results. Test of the 10-effect MEDS without heat insulation and heat recovery shows that the heat leakage reduces R_o by 20 ~ 30%. The 20-effect MEDS test with heat insulation and heat recovery shows that the measured R_o is identical with the estimated result. The other conclusion from the test is that, a single unit of MEDS can produce pure water at 5 L/day (6 hours) at most.

Table 5. MEDS test using electric steam generator

Vapour input, [g/h]	Yield, [g/h]		Measured Ro	Estimated Ro
	Pure water	Salt brine		
10-effect (without heat insulation and heat recovery)				
234	1032	228	4.4	5.45
276	972	744	3.5	4.35
336	960	960	2.9	3.98
20-effect (with heat insulation, without heat recovery)				
90	600	180	6.7	7.35
84	546	240	6.5	6.74
78	480	240	6.1	6.5
20-effect (with heat insulation and heat recovery)				
102	738	624	7.2	7.2
63	492	468	7.8	7.8

The efficiency of the heat recovery device has been tested. Figure 6 shows that the input cold water is preheated and output hot water is cooled. The temperature difference between preheated water and cooled heat is about 8 °C. In other words, the effectiveness of the heat recovery device is greater than 0.9.

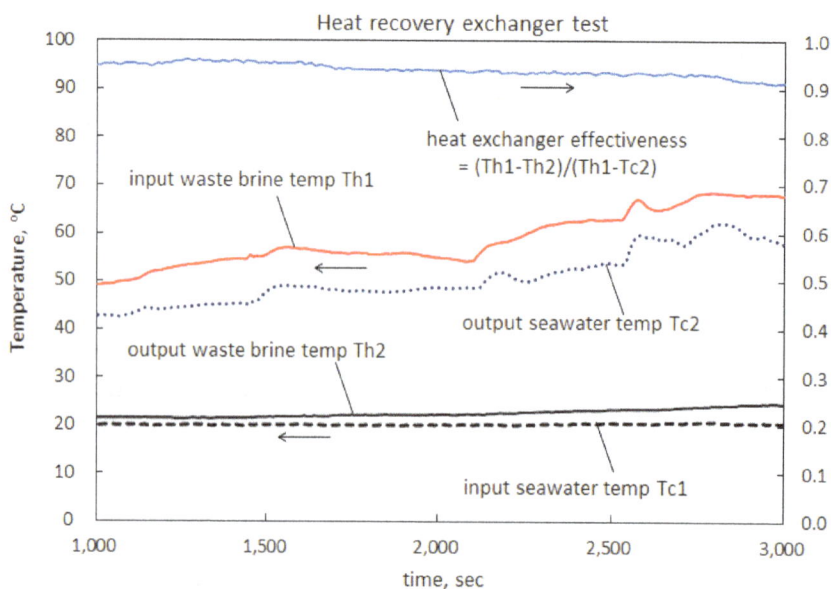

Figure 6. Heat recovery test

Since the spacing of still cell is very small (<6 mm), the condensed pure water flowing down the plate may be contaminated by the seawater across the small gap. We performed a test using dyed seawater in red colour as the input and observe the colour of the produced pure water. No contamination was found.

OUTDOOR PERFORMANCE TEST OF SADS

The prototype of SADS is completed by integrating MEDS with VTSC module (Figure 7). Some outdoor tests with different number of VTSC tubes were carried out.

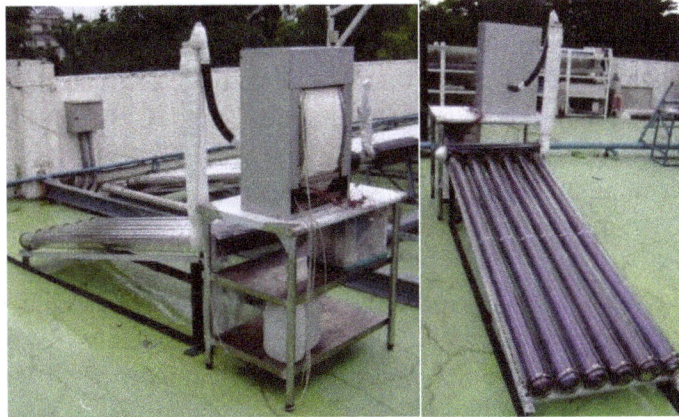

Figure 7. Outlook of SADS

Table 6 shows that at solar radiation >900 W/m^2, one set of MEDS coupled with one tube of VTSC performs at the highest efficiency and the pure water production rate is >18 L/m^2/6 h. This value is close to or even higher than the previous record for similar stills in the literature [8]. At lower solar radiation, more tubes of VTSC can be used to achieve high efficiency. The test has shown that the size of MEDS needs to be enlarged. The measured R_o is good but R_s does not compare to estimation. This indicates that there is a large energy loss for MEDS heat input from the vacuum-tube collector. The data analysis shows that the energy loss is 50-70%, due to vapour leakage from the inlet vapour collector of MEDS. This means that the pure water production rate can increase about 3 fold, if the heat loss is controlled to <20%.

Table 6. SADS test with different number of VTSC tubes

Number of VTSC tubes	Solar Irradiation, [W/m^2]	Production rate, [L/m^2/6 h]	R_o	R_s
3	600	~10	~7	~1.5
2	900	~15	~7	~1.7
1	900	>18	~8	~2

DISCUSSION AND CONCLUSION

The present study intends to develop high-performance solar-assisted desalination technology (SADS) using multiple-effect diffusion-type still (MEDS). The MEDS is designed using an innovative configuration. Performance analysis of MEDS was carried out. A MEDS prototype was built and tested with vacuum-tube solar collector. It shows that the measured result is very close to the estimation. Therefore, 10-effect MEDS will produce pure water at around 12.6 L/day/set (13.7 L/m^2/day) in Taiwan with solar irradiation 600 W/m^2 and 18.1 L/day/set (19.7 L/m^2/day) in desert area with solar irradiation 800 W/m^2.

For 20-effect MEDS, the daily pure water production rate (for 6 h) is around 16.5 L/day/set (17.9 L/m^2/day) in Taiwan and 23.7 L/day/set (25.8 L/m^2/day) in desert area. For 20-effect still, the yield rate increases by 32% compared to 10-effect still.

A design of SADS has been completed by integrating MEDS with VTSC module. At solar irradiation 900 W/m^2, one set of MEDS coupled with one tube of VTSC will perform at the highest efficiency, at pure water production rate >18L/m^2/6 h. The test has shown that the size of MEDS needs to be enlarged, and there is a large input energy loss at MEDS for the heat supply from the vacuum tube collector. The energy loss is 50-70%. The pure water production rate will increase about 3 fold, if the heat loss is controlled to be 20%.

Further experiments on a full-scale MEDS prototype run with solar collector are needed. To achieve the above theoretical target, design modifications of the MEDS are required.

The heat and mass transfer through the small gap inside the still cell involves complicated transport phenomena. The thermal resistances from heating plate conduction, drop-wise condensation, and vapour diffusion process [9] need to be reduced by using various innovative designs. System design matching is also very important. Size matching between the vapour generator and the MEDS units needs more research.

How to keep a reasonable temperature gradient along the path of heat diffusion process is another key factor to achieve good results. Research on the optimum number of multi-effects is also needed.

The polycarbonate heating plate in the handmade unit is 0.5 mm thick. It can be reduced using moulding technique to reduce the conduction thermal resistance. The structural design of still cell can be further simplified for easy assembly and disassembly. Water makeup flow control is also very important in order to maintain good heat and mass transfer, since solar radiation is not steady. An optimal control may be needed.

ACKNOWLEDGEMENT

This publication is based on work supported by Award No.KUK-C1-014-12, made by King Abdullah University of Science and Technology (KAUST), Saudi Arabia.

NOMENCLATURE

C_e	-	salt concentration at exit (brine) (0.36)
C_i	-	salt concentration at the inlet flow (0.036)
c_p	[kJ kg^{-1}K^{-1}]	specific heat of water
h_{fg}	[kJ kg]	latent heat of water
m_e	[kg s^{-1}]	mass flow rate at exit (brine)
m_i	[kg s^{-1}]	total mass flow rate at the inlet
m_v	[kg s^{-1}]	mass flow rate of the evaporated vapour
q_i	[W]	heat input
r	-	ratio of pure water vapour flow to the brine flow
R_s	-	solar energy coefficient of performance (total vapour condensation heat required for actual pure water production over total solar energy input to VTSC), Eqn (7)

| R_o | - | distillation coefficient of performance (total vapour condensation heat required for actual pure water production over total heat input to MEDS), Eqn (8) |
| ΔT | [°C] | temperature difference between input seawater and output salt brine |

REFERENCES

1. Kaushal, V. A., Solar stills: A review. *Renewable and Sustainable Energy Reviews*, 14, pp 446–453, 2010.
2. Cooper, P.I., Appleyard, J.A., The construction and performance of a three effect, wick-type, tilted solar still, *Sun at Work*, 12, pp 4-8, 1967
3. Bouchekima, B. Gros, B., Ouahes, R., Diboun, M., Performance study of the capillary film solar distiller, *Desalination*, 116, pp 185-192, 1998.

4. Tanaka, H., Nosoko, T., Nagata,T., A highly productive basin - type - multiple - effect coupled solar still, *Desalination*, 130, pp 279-293, 2000.

5. Tanaka, H., Nosoko, T., Nagata, T., Parametric investigation of a basin - type multiple - effect coupled solar still, *Desalination*, 130, pp 295-304, 2000.

6. Tanaka, H., Nosoko, T., Nagata,T., Experimental study of basin-type, multiple-effect, diffusion-coupled solar still, *Desalination*, 150, pp 131-144, 2002.

7. Tanaka, H., Nakatake, Y., A vertical multiple-effect diffusion-type solar still coupled with a heat-pipe solar collector, *Desalination*, 160, pp 195-205, 2004.

8. Tanaka, H., Experimental study of vertical multiple-effect diffusion solar still coupled with a flat plate reflector, *Desalination*, 249, pp 34–40, 2009.

9. Rose, J.W., Dropwise condensation theory and experiment: a review, *Proc Inst Mech Engrs Part A: J. Power and Energy*, 216, pp 115-127, 2002.

5

Appliance Diffusion Model for Energy Efficiency Standards and Labeling Evaluation in the Capital of Lao PDR

Hajime Sasaki[*1], *Ichiro Sakata*[2], *Weerin Wangjiraniran*[3],
Sengprasong Phrakonkham[4]

[1]Policy Alternatives Research Institute, The University of Tokyo, 7-3-1, Hongo, Bunkyo-ku, Tokyo, Japan
e-mail: sasaki@pari.u-tokyo.ac.jp
[2]Policy Alternatives Research Institute, The University of Tokyo, 7-3-1, Hongo, Bunkyo-ku, Tokyo, Japan
e-mail: isakata@ipr-ctr.t.u-tokyo.ac.jp
[3]Energy Research Institute, Chulalongkorn University, 12th Floor Institute Building III, Phyathai Road,
Pratumwan, Bangkok, Thailand
e-mail: weerin@eri.chula.ac.th
[4]Faculty of Engineering, National University of Laos, Don Noun, Vientiane, Lao PDR
e-mail: sengprasong@yahoo.com

ABSTRACT

Because of the rapid growth of energy demand in developing countries, policies for energy efficiency are receiving increasing attention. Although Energy Efficiency Standards and Labeling (EES&L) is a standard policy tool in many countries, some developing countries, such as Lao PDR, have not yet implemented them fully. In order to understand the potential impact of EES&L, this paper aims at collecting data that contribute to EES&L and at analysing appliance possessions in Vientiane City, Lao PDR. We conducted an interview survey on 600 households in Vientiane City and performed logistic regression analysis that set possession of appliances as the dependent variable. As a result of the analysis, we identified that the income level and the electricity consumption are the principal independent variables and the relationship of these variables with possession rates depends on appliances. Our model helps identify appliances that are expected to be in high demand associated with either economic growth or human population increase in Vientiane City.

KEYWORDS

Energy efficiency, Standard and labeling, Minimum Energy Performance Standards (MEPS), Possession of appliances, Developing country, Regression analysis.

INTRODUCTION

The economic growth and accompanying industrialization and rising standards of living in developing countries have led to a rapid growth of energy demand. Although Energy Efficient Standards and Labeling (EES&L) for appliances is now a standard policy tool in many countries [1], some developing countries, such as Lao People's Democratic Republic (Lao PDR), are yet to implement them fully. The country has enormous hydropower potential, but energy-efficiency economic growth would allow for more orderly development of such renewable resources.

Energy efficiency labels are assigned to appliances and they describe the energy performance of a product. These labels enable consumers to make informed purchase decisions by communicating to consumer's energy usage and costs associated with a

---

[*] Corresponding author

product [2]. If implemented appropriately, the experience in various countries demonstrates that appliance standards and labels can provide concrete benefits of energy efficiency.

Another policy scheme is a Minimum Energy Performance Standard (MEPS), which literally sets the minimum required energy standard of marketed energy-consuming products. It is now widely employed in the European countries [3].

In a market with an energy-labeling program, all appliances carry an energy rating, allowing those consumers who care about the environment to choose a model that saves energy. By introducing a labeling program, policymakers can induce a market for consumers and citizens who care about the environment and wish to save energy. This mechanism is called "market pull" and increases the average efficiency of models on the market somewhat, as shown in Figure 1. In addition to the labeling system, MEPS can be introduced, which disables the sale of products of the most inefficient kind. MEPS provides a "market push," which also increases the average efficiency in the market. The two kinds of policies, energy labeling and MEPS, can be used to improve the efficiency of marketed products [4].

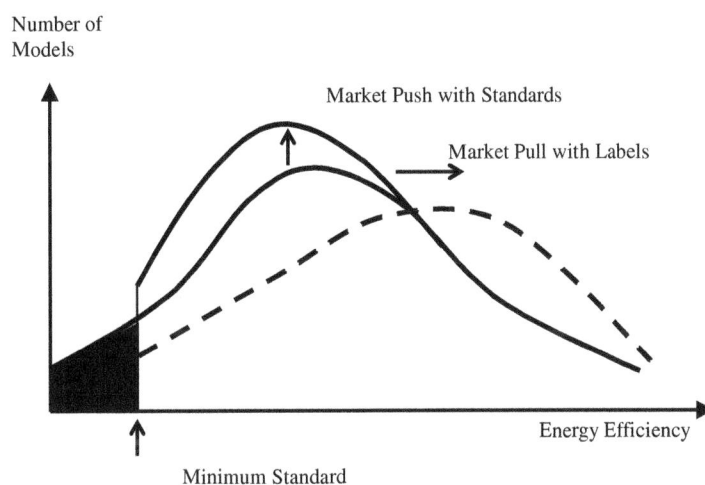

Figure 1. The effect of MEPS and labeling on the efficiency distribution of the products
(Source: CLASP, 2005 [2] edited by author)

Existing research on EES&L is mostly limited to developed and emerging economies. Rosenquist *et al.* [5] analysed impacts of possible improvements in energy efficiency standards for residential and commercial sectors in the US, concluding that such improvements may provide substantial benefits. Augustus de Melo and de Martino Jannuzzi [6] developed a model to assess the impacts of MEPS (particularly refrigerators) in Brazil, where the MEPS policy is in its early stage. Japan has been implementing the Top Runner Standard for various appliances, which is similar to a MEPS scheme. According to an estimate, this schemed achieved energy savings of 55.2% relative to the refrigerators shipped in 1998 [7]. Korea fervently utilized a mix of mandatory energy labeling and MEPS to improve the appliance efficiencies in the 1990s. They achieved a rather astonishing result. In a matter of 7 years, they saw a 38% improvement in the average efficiency of fluorescent lamps. Likewise, the efficiency of refrigerator-freezers is increased by 42% [8]. CLASP (Collaborative Labeling and Appliance Standards Program) explored the benefits of China's existing (as opposed to additional) standards and labeling [9]. The EES&L programs already implemented in China are expected to lead to a cumulative saving of 1,143 TWh by 2020.

Summary of electrification situation in Lao PDR

Lao PDR's economy is growing steadily in an increasingly sustainable way, with the reforms underway reducing poverty. Its real GDP grew by 7.1% per year on average for 2000-2010. In 2010, the Gross National Income (GNI) per capita of Lao PDR reached USD 1,040 and, the country is no longer a lower economy income country; it is now a lower-middle income economy. At the pace of current economic growth, Lao PDR's long-term vision is within reach, and it should be able to graduate from Least Developed Country status by 2020.

Accompanying this rapid economic growth is a fast increase in energy demand in all sectors. According to the analysis by the Institute of Energy Economics of Japan, the transportation sector is expected to grow five-fold from the current level of consumption. The electricity demand has been and is also expected to rise substantially. Figure 2 shows that the average growth rate of electricity consumption for 2000-2010 is 11% [10]. Moreover, Lao PDR has been electrifying rural areas successfully (Figure 3), with an ambition to achieve an electrification rate of over 90% in 2020. This ambitious plan will be achieved by massively extending the power grid even to peripheral areas.

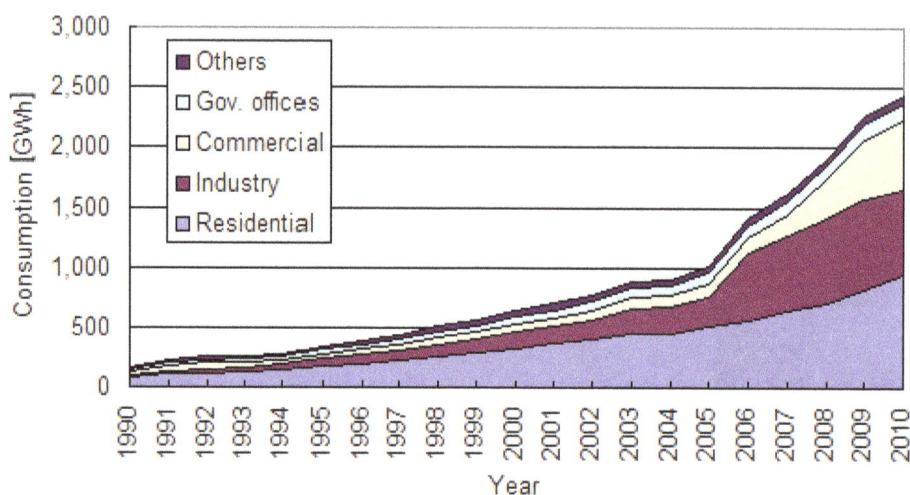

Figure 2. Total consumption of electric power
(Source: Department of Electricity, Ministry of Energy and Mines in Lao PDR, 2010 [10])

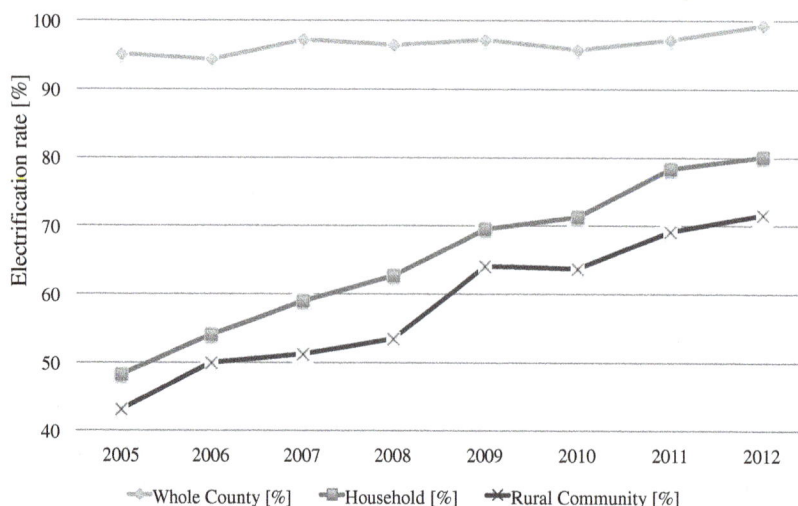

Figure 3. The electrification rate in Lao PDR

Sasaki *et al.* [11] proposed a roadmap scenario of electricity technologies for Lao PDR and they conducted stakeholder meetings to foster consensus building with the Lao PDR government. However, the study emphasized the supply side over the demand efficiency. A deeper analysis from the demand perspective is urgently necessary. For instance, they will likely need to implement EES&L in the near future, and there is a need for policy discussions on the potential EES&L design. A model for EES&L evaluation would be helpful to facilitate the policy design process. According to Michael *et al.* an econometric model of appliance diffusion serves as the basis for subsequent analyses since the future energy efficiency distribution of appliances is the starting point for assessing energy efficiency policies. Projections based on such models [12] can analyse the impact of the main drivers (wealth, urbanization, and electrification).

The purpose of this paper is to develop an econometric model of appliance diffusion that can serve as the basis for evaluation of EES&L policies for Vientiane City, Lao PDR. As households become richer, they are expected to possess more and more appliances, with the income being a key indicator. On the other hand, different appliances should diffuse at the different levels of income, and some might even saturate early in the development stage. This motivates a model that takes into account multiple drivers associated with economic growth and socio-economic changes.

We therefore investigate which key driver contributes to the diffusion rate of each appliance in Vientiane City. Specifically, we implement a logistic regression analysis that combines income, household size and district. These results show this appliance is likely to diffuse early in the economic development allowing for priority setting in the EES&L policy. Ideally such a regression analysis should be based on comprehensive, public statistics. Unfortunately, developing economies such as Lao PDR do not maintain data sufficient for the present purpose of demand-side analysis. We thus conducted a survey that collected the data on household possession of appliances in Vientiane City. To our knowledge, this is the first such data collection exercise. Our regression analyses are based on the collected data.

METHODOLOGY

We conducted an interview survey that mainly focused on the possession of appliances in a residential sector. In this survey, Vientiane, the capital city of Lao PDR, is set as the target area. We randomly sampled from the capital, which is divided into 5 districts and 31 villages.

The total population in Vientiane city is about 980,000 in 2013. Based on an estimation of sample size we assumed 600 households interview is enough for estimation in this city.

In this paper, we implemented a logistic regression analysis that explains the possessions of appliances as dependent variable. The appliances of our focus are refrigerators, air-conditioners, and TVs (CRT and LCD/LED), because these products are often targeted by EES&L policies in other countries. For independent variables, we have the following:

- Income level dummy;
  - 0 – 2,000,000 kip/month;
  - 2,000,001 – 4,000,000 kip/month;
  - 4,000,001 – 6,000,000 kip/month;
  - 6,000,000 – 8,000,000 kip/month;
  - 8,000,001 – 10,000,000 kip/month;
  - 10,000,001 – 12,000,000 kip/month;
  - 10,000,001 – 14,000,000 kip/month;

○   10,000,001 – 16,000,000 kip/month;
○   Or 10,000,001–18,000,000 kip/month.

The number of males, the number of females, monthly consumption of electricity kWh/month, type of house dummy (concrete, wooden or mixed) and district dummy (Hadsaifong district, Saysettha district, Sikhotabong district or Sisatthanak district).

RESULTS

Summary of the data

By two-stage random sampling, we approached 600 households in Vientiane City, and obtained 577 responses. Figure 4 shows a histogram of the monthly household income answered by 577 subjects. The modal class is 2,000,000 kip (260 USD)/month – 4,000,000 kip (520 USD)/month and accounts for 40% of the total. So we can assume that this demographic class is the middle-income bracket in Vientiane City. Similarly, Figures 5 and 6 describe histograms of the monthly consumption of electricity kWh/month and the monthly average electricity cost kip/month, respectively. Most respondents are using below 500 kWh/month. On the other hand, some households exceed 3,000 kWh/month. As there is a strong correlation between the monthly consumption of electricity and the monthly average electricity cost (Adjusted R-squared: 0.82), we cannot use both of them as independent variables; otherwise, we would face a problem of multicollinearity in the regression analysis. In the following, we select the monthly consumption of electricity as an independent variable.

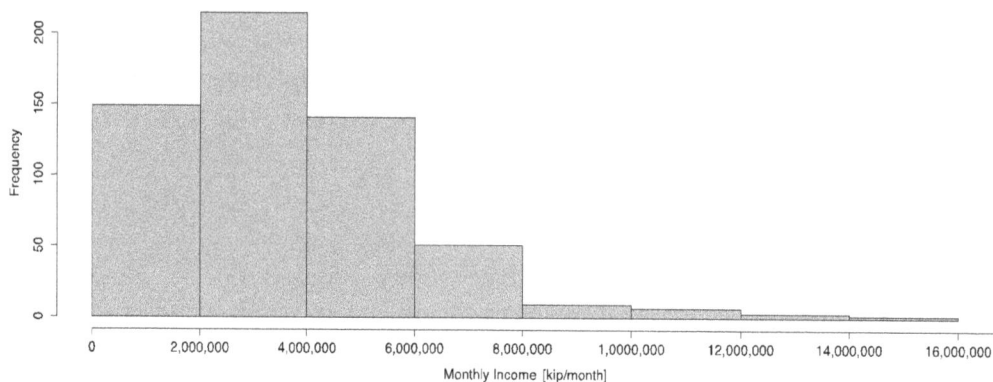

Figure 4. Histogram of monthly income [kip/month]

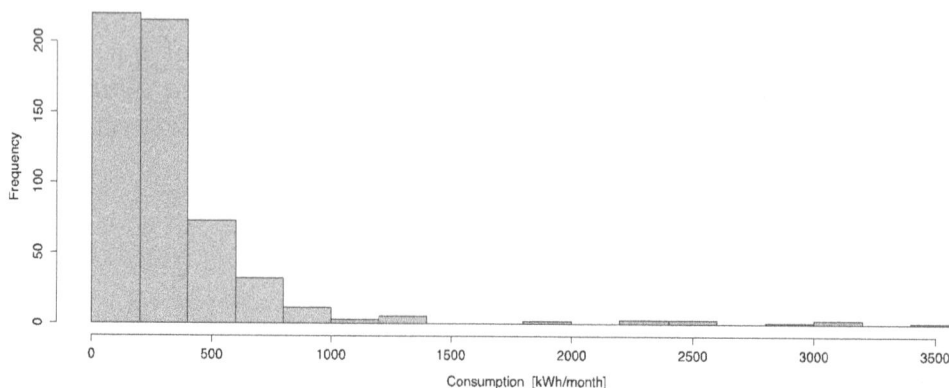

Figure 5. Histogram of monthly consumption of electricity [kWh/month]

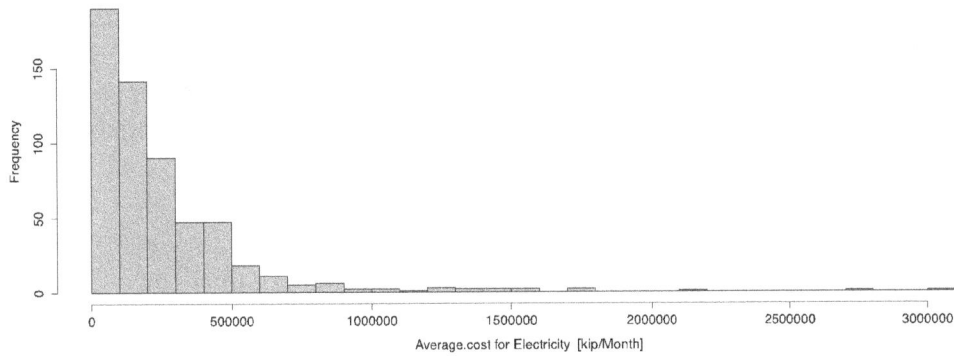

Figure 6. Histogram of monthly average electricity cost [kip/month]

Figure 7 shows the number of households that possess each appliance (by the bar chart with left axis) and the average number of units of each appliance in each household (by the line chart with right axis). As can be seen from the bar chart with the left axis, there are very few households that do not have mobile phones (household diffusion rate of 98%). According to "World Telecommunication/ICT Indicators Database, June 2012", the diffusion rate of mobile phones was 87.2% in the whole of Lao PDR in 2011. For other appliances, the diffusion rates are as follows:

- Refrigerators 94%;
- Air conditioners (Split wall type) 50%;
- Air conditioners (Split floor type) 7%;
- TVs (CRT) 83%;
- TVs (LCD/LED) 38%;
- CFLs 5%,
- Vacuum machines 5%;
- PCs 17%;
- Microwaves 25%.

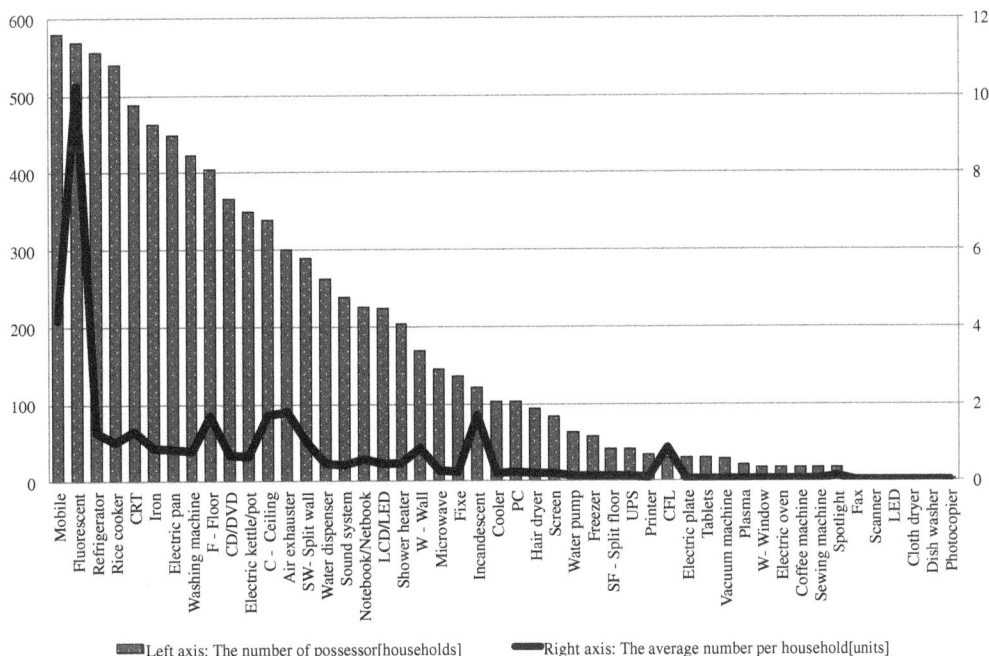

Figure 7. The number of possessors (households), (left axis: bar chart); and the average number in households (units), (right axis: line chart)

On the other hand, the line chart with right axis shows the number obtained from the total number of ownership respondents of each appliance divided by the number of respondents from 577 households. For example, the average number of units of fluorescent lamps in each household is 10 units. In contrast, the average number of units of mobile phones is 4.

Regression analysis

Based on these data we conducted logistic regression in which the dependent variable is possession (binary data) of:

- Refrigerators;
- Air-conditioners (Split wall type and Split floor type),
- TVs (CRT type and LCD/LED type);
- CFLs;
- Vacuum machines;
- PCs;
- Microwaves.

Table 1 indicates the result of logistic regression analysis in which the dependent variable is possession of a refrigerator. The significant contribution of income level to possession of a refrigerator is mentioned in the Bottom-Up Energy Analysis System model (BUENAS model) as EES&L forecasting model [1]. From our result, any independent variables do not contribute to possession of refrigerators. Because the diffusion rate of refrigerators in this city is 94% as explained in Figure 7, it is quite natural that it should be so and we can understand that refrigerators are universally distributed regardless of income level, the size and type of family and geo-location.

Table 1. The result for refrigerators

	Estimate	Std. error	z value	Pr ($>|z|$)
(Intercept)	1.72E+01	3.96E+03	0.004	0.997
Monthly income	8.58E-02	1.32E-01	0.651	0.515
The number of male	8.72E-02	1.52E-01	0.574	0.566
The number of female	7.14E-02	1.37E-01	0.52	0.603
Electric consumption	2.05E-04	5.51E-04	0.371	0.71
Concrete house	-1.50E+01	3.96E+03	-0.004	0.997
Wooden house	-9.53E-02	4.04E+03	0	1
Mixed house	-1.49E+01	3.96E+03	-0.004	0.997
Hadsaifong district	5.32E-01	6.68E-01	0.798	0.425
Saysettha district	-3.72E-01	5.55E-01	-0.671	0.502
Sikhotabong district	3.33E-01	6.23E-01	0.541	0.588
Sisatthanak district	-2.93E-01	5.84E-01	-0.502	0.615

--- Signif. Codes: 0 '***' 0.001 '**' 0.01 '*' 0.05 '.' 0.1 ' ' 1

Table 2 indicates the result of analysis in which the dependent variable is possession of air-conditioners (Split wall type). Table 3 shows the result of analysis in which the dependent variable is possession of air-conditioners (Split floor type). From the Table 2, we can see that the monthly income level and electricity consumption contribute significantly ($P<0.001$). We should interpret that there is causality between the possession of air-conditioner and the monthly income and electricity consumption. It is implicit that the households that use air-conditioners (Split wall type) consume a large amount of electricity. It is interesting that the variable of "Sisatthanak district" strongly contributes to the possession of air-conditioners (Split wall type), meaning that this type of air conditioners is highly distributed in this district. On the other hand, from the Table 3, we can figure out that any independent variables do not contribute to the possession of air-conditioners (Split wall type) without "Saysettha district" variable. From this result, the diffusion of the type of air conditioners depends on the district in Vientiane city.

Table 2. The result of air-conditioners (Split wall type)

	Estimate	Std. error	z value	Pr $(>|z|)$
(Intercept)	-1.46E+01	5.35E+02	-0.027	0.9782
Monthly income	3.72E-01	7.41E-02	5.017	5.26E-07***
The number of male	6.31E-02	7.42E-02	0.85	0.3955
The number of female	9.62E-03	6.07E-02	0.158	0.8742
Electric consumption	3.23E-03	5.63E-04	5.739	9.50E-09***
Concrete house	1.29E+01	5.35E+02	0.024	0.9808
Wooden house	1.25E+01	5.35E+02	0.023	0.9813
Mixed house	1.27E+01	5.35E+02	0.024	0.9811
Hadsaifong district	8.42E-02	3.04E-01	0.277	0.7814
Saysettha district	5.02E-02	2.96E-01	0.17	0.8652
Sikhotabong district	-2.87E-01	2.93E-01	-0.98	0.3269
Sisatthanak district	-6.49E-01	3.18E-01	-2.037	0.0417*

--- Signif. Codes: 0 '***' 0.001 '**' 0.01 '*' 0.05 '.' 0.1 ' ' 1

Table 3. The result of air-conditioners (Split floor type)

	Estimate	Std. error	z value	Pr $(>|z|)$
(Intercept)	-1.46E+01	8.83E+02	-0.017	0.9868
Monthly income	9.46E-02	1.15E-01	0.823	0.4105
The number of male	-1.13E-02	1.41E-01	-0.08	0.9361
The number of female	-1.42E-01	1.32E-01	-1.078	0.2808
Electric consumption	6.95E-04	4.29E-04	1.618	0.1056
Concrete house	1.24E+01	8.83E+02	0.014	0.9888
Wooden house	1.28E+01	8.83E+02	0.015	0.9884
Mixed house	1.25E+01	8.83E+02	0.014	0.9887
Hadsaifong district	-5.47E-01	5.18E-01	-1.057	0.2904
Saysettha district	-2.97E+00	1.09E+00	-2.732	0.0063**
Sikhotabong district	-6.11E-01	5.15E-01	-1.185	0.2359
Sisatthanak district	6.59E-02	4.72E-01	0.139	0.8891

--- Signif. Codes: 0 '***' 0.001 '**' 0.01 '*' 0.05 '.' 0.1 ' ' 1

Table 4 indicates the result of analysis in which the dependent variable is possession of TV (CRT). Table 5 indicates the result of analysis in which the dependent variable is possession of a TVs (LCD/LED). The possession of TVs (CRT) is highly contributed by some variables; the monthly income ($P<0.1$), the number of women in the household

($P<0.05$), electric consumption ($P<0.1$), Sikhotabong district ($P<0.01$) and Sisatthanak district ($P<0.05$). Since the diffusion rate of TVs (CRT) is 83%, it is estimated that the appliances have already been widely diffused to the public. According to the Table 5, significant variables of possession of TVs (LCD/LED) are monthly income ($P<0.001$), electric consumption ($P<0.001$) and Sikhotabong district ($P<0.01$). The number of households that possess an LCD/LED is 223 and the diffusion rate of LCD/LED is 38%. In general, LCD/LED-TVs belong to the expensive category in televisions, so it is reasonable that income level contributes significantly.

Table 4. The result of TVs (CRT)

	Estimate	Std. error	z value	Pr (>|z|)
(Intercept)	1.24E+01	5.35E+02	0.023	0.9815
Monthly income	-1.42E-01	7.68E-02	-1.843	0.06528
The number of male	3.88E-02	9.30E-02	0.417	0.67653
The number of female	2.29E-01	9.42E-02	2.435	0.0149*
Electric consumption	-4.57E-04	2.61E-04	-1.749	0.08022
Concrete house	-1.17E+01	5.35E+02	-0.022	0.98263
Wooden house	-1.04E+01	5.35E+02	-0.019	0.98452
Mixed house	-1.18E+01	5.35E+02	-0.022	0.98249
Hadsaifong district	5.21E-01	3.41E-01	1.526	0.127
Saysettha district	5.39E-01	3.31E-01	1.631	0.10296
Sikhotabong district	0.65E-01	3.68E-01	2.624	0.00869**
Sisatthanak district	9.19E-01	3.91E-01	2.349	0.01883*

--- Signif. Codes: 0 '***' 0.001 '**' 0.01 '*' 0.05 '.' 0.1 ' ' 1

Table 5. The result of TVs (LCD/LED)

	Estimate	Std. error	z value	Pr (>|z|)
(Intercept)	-1.40E+01	5.35E+02	-0.026	0.9791
Monthly income	2.42E-01	6.65E-02	3.643	0.00027***
The number of male	3.43E-02	7.36E-02	0.464	0.64249
The number of female	-6.71E-02	6.41E-02	-1.048	0.29482
Electric consumption	1.42E-03	3.51E-04	4.047	5.19E-05***
Concrete house	1.31E+01	5.35E+02	0.024	0.98046
Wooden house	1.09E+01	5.35E+02	0.02	0.98376
Mixed house	1.28E+01	5.35E+02	0.024	0.98094
Hadsaifong district	-2.53E-01	2.93E-01	-0.862	0.38857
Saysettha district	-4.17E-01	2.87E-01	-1.451	0.1467
Sikhotabong district	-9.18E-01	3.04E-01	-3.018	0.00255**
Sisatthanak district	-3.36E-01	3.05E-01	-1.101	0.2708

--- Signif. Codes: 0 '***' 0.001 '**' 0.01 '*' 0.05 '.' 0.1 ' ' 1

Table 6 explains that monthly income contributes for the possessions of CFLs. CFLs are well-known appliances for EES&L in other countries. CFLs are energy efficient but high-price appliances. Then, it is no wonder that the monthly income contributes for the possessions. It is interesting that the number of women in the household also contributes to the possession. Another interesting fact about the diffusion of CFLs is that s appliances are obviously distributed in Saysettha district than other districts: Hadsaifong, Sikhotabong and Sisatthanak. It is to be noted that our analysis does not model the number of each appliance in the households but possession (yes or no) of each appliance.

Table 6. The result of CFLs

	Estimate	Std. error	z value	Pr (>|z|)
(Intercept)	-1.43E+01	8.83E+02	-0.016	0.987
Monthly income	-3.24E-01	1.46E-01	-2.222	0.0263[*]
The number of male	2.21E-01	1.55E-01	0.142	0.8868
The number of female	1.84E-01	8.87E-02	2.071	0.0384[*]
Electric consumption	3.27E-04	3.51E-04	0.934	0.3504
Concrete house	1.07E+01	8.83E+02	0.012	0.9903
Wooden house	1.07E+01	8.83E+02	0.012	0.9904
Mixed house	1.04E+01	8.83E+02	0.012	0.9906
Hadsaifong district	-4.16E-01	9.34E-01	-0.446	0.6559
Saysettha district	1.63E+00	6.80E-01	2.398	0.0165[*]
Sikhotabong district	-2.98E-01	8.52E-01	-0.349	0.7269
Sisatthanak district	-1.11E-01	8.46E-01	-0.131	0.8961

--- Signif. Codes: 0 '***' 0.001 '**' 0.01 '*' 0.05 ',' 0.1 '' 1

It is difficult to say that monthly income contributes to the possession of vacuum machines, according to the result of analysis as shown in Table 7. On the other hand, the electric consumption contributes strongly to possession. This appliance is widely spread in Hadsaifong district.

Table 7. The result of vacuum machines

	Estimate	Std. error	z value	Pr (>|z|)
(Intercept)	-1.89E+01	1.08E+04	-0.002	0.9986
Monthly income	-1.49E-01	1.33E-01	-1.126	0.2602
The number of male	8.45E-03	1.66E-01	0.051	0.9595
The number of female	-1.04E-01	1.52E-01	-0.687	0.4919
Electric consumption	1.61E-03	3.45E-04	4.672[***]	2.99E-06
Concrete house	1.67E+01	1.08E+04	0.002	0.9988
Wooden house	7.96E-02	1.10E+04	0	1
Mixed house	1.02E-01	1.08E+04	0	1
Hadsaifong district	-3.04E+00	1.27E+00	-2.396	0.0166[*]
Saysettha district	-8.49E-01	6.21E-01	-1.368	0.1713
Sikhotabong district	-7.76E-01	6.51E-01	-1.192	0.2334
Sisatthanak district	-5.26E-01	6.20E-01	-0.849	0.3961

--- Signif. Codes: 0 '***' 0.001 '**' 0.01 '*' 0.05 ',' 0.1 '' 1

Table 8 shows the result of analysis about possession of PCs. This result of PCs is similar with the result for vacuum machines. In both of them, electric consumption contributes strongly but monthly income does not. From this result we can estimate that these appliances are high-consumption electric appliances. We can also assume that PCs may be left turned on the whole day in some households.

Table 8. The result of PCs

	Estimate	Std. error	z value	Pr ($>|z|$)
(Intercept)	-1.42E+01	5.35E+02	-0.026	0.97889
Monthly income	6.13E-02	7.53E-02	0.815	0.41529
The number of male	1.31E-01	8.51E-02	1.542	0.12315
The number of female	4.75E-02	6.42E-02	0.74	0.45931
Electric consumption	6.57E-04	2.47E-04	2.664**	0.00772
Concrete house	1.20E+01	5.35E+02	0.022	0.98208
Wooden house	1.13E+01	5.35E+02	0.021	0.98319
Mixed house	1.17E+02	5.35E+02	0.022	0.98264
Hadsaifong district	-7.11E-01	4.02E-01	-1.767	0.07715
Saysettha district	-5.63E-01	3.72E-01	-1.514	0.13014
Sikhotabong district	2.18E-01	3.40E-01	0.64	0.52189
Sisatthanak district	-2.89E-01	3.81E-01	-0.758	0.4485

--- Signif. Codes: 0 '***' 0.001 '**' 0.01 '*' 0.05 ',' 0.1 ' ' 1

From the Table 9, we can understand that the microwave is a luxury item for Vientiane's citizens. The analysis also reflects that this electric appliance is a high-consuming one.

Table 9. The result of microwaves

	Estimate	Std. error	z value	Pr ($>|z|$)
(Intercept)	-1.43E+01	5.35E+02	-0.027	0.9787
Monthly income	1.78E-01	7.22E-02	2.461	0.0139*
The number of male	-1.80E-02	8.34E-02	-0.216	0.8287
The number of female	2.54E-02	6.32E-02	0.402	0.6878
Electric consumption	1.83E-03	3.86E-04	4.753***	2.01E-06
Concrete house	1.20E+01	5.35E+02	0.022	0.9821
Wooden house	1.17E+01	5.35E+02	0.022	0.9825
Mixed house	1.18E+01	5.34E+02	0.022	0.9824
Hadsaifong district	-2.14E-01	3.68E-01	-0.583	0.5599
Saysettha district	2.84E-01	3.35E-01	0.85	0.3955
Sikhotabong district	3.67E-02	3.50E-01	0.105	0.9164
Sisatthanak district	-3.50E-03	3.63E-01	-0.01	0.9923

-- Signif. Codes: 0 '***' 0.001 '**' 0.01 '*' 0.05 ',' 0.1 ' ' 1

DISCUSSION

As we mentioned in the RESULT part, the diffusion rate of refrigerators is over 94%. It means refrigerators are widely installed in most households in Vientiane city and it is no wonder the possession of refrigerators is not significantly affected by any independent variables. It is also means the impact of EES&L will be highly promising.

According to this paper's co-author Phrakonkham, who lives in Vientiane city, most of energy efficiency labels on the refrigerators that have already diffused are forged. He also said one could import the labels from Thailand without refrigerators itself. This country has to immediately take corrective actions in such a situation.

The tendency of *P*-value relating to district about possession of air-conditioners depends on the types (Split wall type or Split floor type). The possession of air-conditioners (Split wall type) is significantly contributed by Monthly income and Electric consumption. On the other hand, these independent variables do not contribute for the possession of Split floor type air-conditioners. The possession of air-conditioners (Split wall type) is significantly contributed by Sisatthanak district. On the other hand, Split floor type is significantly contributed by Saysettha district. Actually, the diffusion rate is obviously different between these types; Split wall type is installed in 50% of the households and Split floor type is installed only in 7% of households. It means that most air conditioners (Split floor type) are installed in Saysettha district.

About the possession of TVs, the contribution of monthly income and electric consumption to possession of TVs (LCD/LED) is stronger than TVs (CRT). Even though TVs (CRT) is widely installed in Vientiane, the number of shipments will decrease in future. The country should lead to replace such high consuming appliances with other type of TVs (LCD/LED or plasma).

Vacuum machines and PCs are common type of appliances. These possessions are contributed by electric consumption but not by monthly income. We can assume from this result that these appliances are diffused regardless of income level and are highly consuming electric appliances.

In general, it is reasonable to implement EES&L policy for appliances that have high-energy efficiency potential, although it depends on the situation of the country or city. Population growth induces economic growth. Economic growth induces population growth. If the country or city will have high population in the near future, it is reasonable to focus on the appliances that have significant contribution from population. If the country will have high economic growth in the near future, it is reasonable focus on the other appliances, air-conditioners (Split wall) and CFLs.

CONCLUSIONS

Lao PDR is expected to achieve high economic growth and an increase in population. If the country introduces EES&L, it will need data from the demand side, especially about the possession of electric appliances. In this research, we collected these data from Vientiane City and identified some variables that significantly contribute to the possession rate.

We conducted an interview survey on 600 households in Vientiane City, and then we succeeded to collect 577 households' data. We used logistic regression analysis to identify independent variables that significantly contribute to the possession of each appliance.

From the results of our analysis, we identified that the the electric consuming are the principal independent variables and the relationship of these variables with possession rates depends on appliances. And we discussed the model for diffusion of appliances in Vientiane City. Based on the recognition that Lao PDR should implement EES&L as soon as possible alongside rapid economic growth, we contributed to the implication as to which appliances should be prioritized for implementation of EES&L. Because of the limitation of the research, there is still a lot of room for future studies. In the future, we will implement more specific analysis using the data, and present these results and implications to the government of Lao PDR. Ideally, not only the residential sector but also the commercial sector should be considered and surveyed to identify the effective appliances for implementation of EES&L in the whole of Lao PDR in the future.

ACKNOWLEDGEMENT

This research was supported by the ERIA (Economic Research Institute for ASEAN and East Asia). The project name was "Energy Efficiency Roadmap Formulation in East Asia" for ERIA Research Project on "Analysis of Energy Saving Potential in East Asia".

REFERENCES

1. McNeil, M. A., *Global Potential of Energy Efficiency Standards and Labeling Programs*, Lawrence Berkeley National Laboratory, 2008.

2. CLASP, Energy Efficiency Labeling Programs aim to Shift Markets for Energy-using Products toward Improved Energy Efficiency, http://www.clasponline.org/en/WhatWeDo/EnergyEfficiencyLabeling.asp, [Accessed: 01-May-2013]

3. Energy Charter Secretariat, Policies that Work: Introducing Energy Efficiency Standards and Labels for Appliances and Equipment, ISBN: 978-905948-081-0, 2009.

4. Wiel, S., & McMahon, J. E., *Energy-efficiency Labels and Standards: A Guidebook for Appliances, Equipment, and Lighting*, Lawrence Berkeley National Laboratory, 2005.

5. Rosenquist, G., et al., Energy Efficiency Standards for Equipment: Additional Opportunities in the Residential and Commercial Sectors, *Energy Policy,* Vol. 34, No. 17, pp 3257-3267, 2006.

6. de Melo, C. A., & Jannuzzi, G. M., Energy Efficiency Standards for Refrigerators in Brazil: A Methodology for Impact evaluation, *Energy Policy,* Vol. 38, No. 11, pp 6545-6550, 2010.

7. Murakami, S., et al., *Energy Consumption, Efficiency, Conservation, and Greenhouse Gas Mitigation in Japan's Building Sector*, Lawrence Berkeley National Laboratory, 2006.

8. Lee, S-K.,, MEPS Experience in Korea, Lessons Learned in Asia: Regional Conference on Energy Efficiency Standards and Labeling, Bangkok, 2001

9. Fridley, D., et al., *Impacts of China's Current Appliance Standards and Labeling Program to 2020*, Lawrence Berkeley National Laboratory, 2007.

10. Department of Electricity, Ministry of Energy and Mines in Lao PDR, ELECTRICITY STATISTICS YEARBOOK OF LAO PDR, 2010.

11. Sasaki, H., et al. Energy Efficiency Road Mapping in Three Future Scenarios for Lao PDR, *Journal of Sustainable Development of Energy, Water and Environment Systems,* Vol.1, No. 3, pp 172-186, 2013.

12. McNeil, M. A., & Letschert, V. E., Modeling diffusion of Electrical Appliances in the Residential Sector, *Energy and Buildings*, Vol. 42, No. 6, pp 783-790, 2010.

6

The Role of Bioenergy in Ireland's Low Carbon Future – is it Sustainable?

Alessandro Chiodi[*1], *Paul Deane*[1], *Maurizio Gargiulo*[2], *Brian Ó Gallachóir*[1]

[1]Environment Research Institute, University College Cork, Cork, Ireland
e-mail: a.chiodi@ucc.ie
[2]E4SMA, S.r.l., Torino, Italy

ABSTRACT

This paper assesses through scenario analysis the future role of bioenergy in a deep mitigation context. We focus in particular on the implications for sustainability – namely, competing demands for land-use, import dependency, availability of sustainable bioenergy and economics. The analysis here is limited to one Member State, Ireland, which is an interesting case study for a number of reasons, including significant import dependency and recent acceleration in renewable energy deployment. We used the Irish TIMES model, the energy systems model for Ireland developed with the TIMES model generator, for this scenario analysis. Long term, least cost mitigation scenarios point to bioenergy meeting more than half of Ireland's energy needs by 2050. The results of this paper point to the impact of tightened sustainability criteria and limitation on bioenergy imports, namely the increased use of indigenous bioenergy feedstocks, increased electrification in the energy system, the introduction of hydrogen and higher marginal abatement costs.

KEYWORDS

Bioenergy, Sustainability, Emissions mitigation, Climate policy, Energy systems modelling, MARKAL-TIMES.

INTRODUCTION

Due to growing worldwide concerns regarding anthropogenic interference with the climate system, 141 countries have, since December 2009, associated themselves with the Copenhagen Accord [1] that declared that deep cuts in Global Greenhouse Gas (GHG) emissions are required so as to hold the increase in global temperature below 2 degrees Celsius. Despite recent projections [2] indicate that the world is not on track to meet this 2 °C target – the long-term average temperature increase is more likely to be between 3.6 °C and 5.3 °C – it remains technically feasible, though extremely challenging [3]. To keep open a realistic chance of meeting the 2 °C target, intensive action is required before 2020, the date by which a new international climate agreement is due to come into force.

The European Union (EU) perspective is that industrialized countries should contribute to this global emissions reduction target by reducing GHG emissions by 20% by the year 2020 and between 80% and 95% by the year 2050, relative to 1990 levels. Even in the absence of a wider international agreement on climate policy, the EU has set an ambitious climate and energy policy framework for 2020 [4-7] and is now reflecting

---

[*] Corresponding author

on a new 2030 framework [8]. Moreover European Commission (EC) laid out long term roadmaps which commit for reductions between 80% and 95% by 2050 relative to 1990 levels [9-11]. Table 1 illustrates the EC perspective on how the mitigation target should be distributed amongst sectors [9].

Table 1. EU Low Carbon Roadmap GHG reduction compared to 1990

Sectors	2005	2030	2050
Power (CO_2)	-7%	-54 to 68%	-93 to -99%
Industry (CO_2)	-20%	-34 to -40%	-83 to -87%
Transport (incl. CO_2 aviation, excl. maritime)	30%	+20 to -9%	-54 to -67%
Residential and services (CO_2)	-12%	-37 to -53%	-88 to -91%
Agriculture (non-CO_2)	-20%	-36 to -37%	-42 to -49%
Other non-CO_2 emissions	-30%	-72 to -73%	-70 to -78%
Total	-7%	-40 to -44%	-79 to -82%

This paper focuses on Ireland, which is an interesting case study relative to other EU Member States (MS). Firstly in Ireland despite the recent economic recession, energy demand growth over the period 1990 to 2011 has been significant (1.8% per annum on average [12]) driven largely by high economic growth (4.8% per annum on average growth in real GDP). This increased energy demand was supplied mainly by fossil fuels, which accounted for 94% of all primary energy used in Ireland in 2011. Oil is the dominant energy source with a share of 49% in 2009 (was 47% in 1990), followed by natural gas with a share of 30% and coal (9%). Renewable energy passed from a low base of 1.8% of primary energy requirement, to 6%, largely driven by increase in wind energy capacity [12]. The rapidly increasing consumption of energy in Ireland, combined with the decreasing domestic production, has resulted in a significant increase in energy imports in recent years. Ireland exhibits a significant dependence on imported fossil fuels, which accounted for 88% in 2011 [12]. The UK is the major source of oil and natural gas for Ireland [13, 14]. Moreover this resulted in a 25.3% growth in energy-related carbon dioxide (CO_2) levels for the period while EU emissions declined [15], as showed in Figure 1. If we reference GHG emissions reductions against 1990 levels rather than 2010 levels results in a very different scale of challenge: an 80% emissions reduction target relative to 1990 levels is equivalent in Ireland to an 82% emissions reduction target relative to 2010 levels, while for EU is 76%.

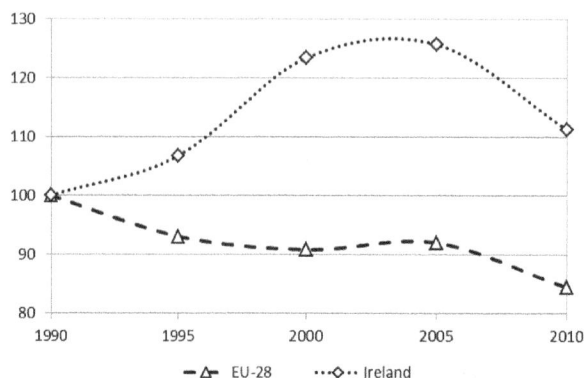

Figure 1. Historical GHG Emissions in EU-28 and in Ireland indexed to 1990

The second distinguishing characteristic of Ireland is the importance of the agricultural sector in the energy and climate debate. Agriculture in Ireland is predominantly based on dairy and beef production from ruminant animals, most of which (over 80%) is exported. Livestock activities are largely based on extensive, grass-based farming. Approximately 82% of total agricultural area in 2010 is devoted to grass (silage, hay and pasture), while the remainder is allocated to rough grazing (11%) and crop production (7%) [16]. In terms of the total land area of Ireland, agriculture accounts for about 60% as shown in Figure 2.

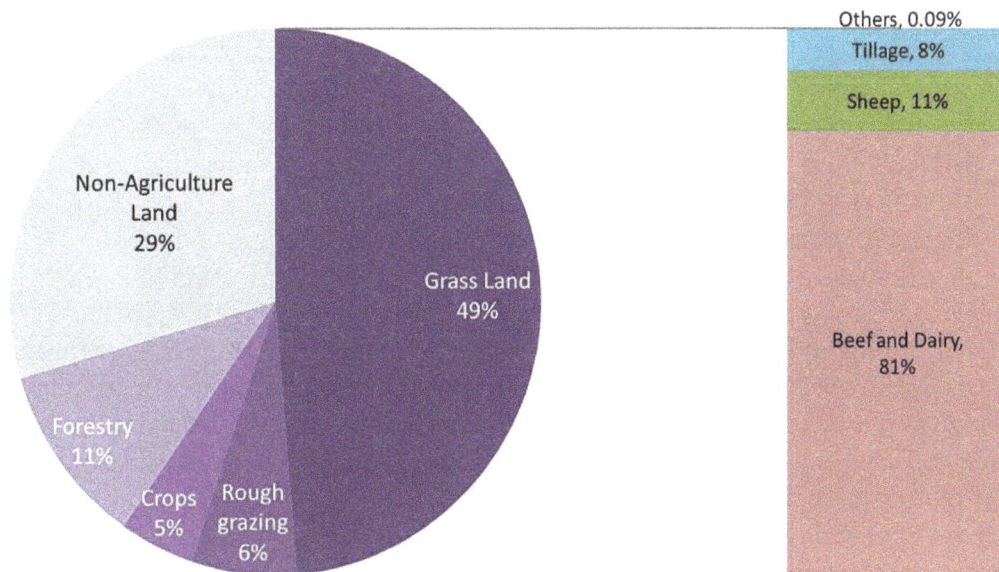

Figure 2. Breakdown of land use in Ireland in 2010

The agri-food sector contributes approximately 7% to Ireland's economy (in terms of GDP), but at the same time agriculture accounts for 32.1% (in 2011) of total GHG compared with just 11.9% for the EU (average across EU-28) [15]. Of these emissions only 5% are associated with energy (for combustion) while the remaining originates as non-combustion emissions (namely methane and nitrous oxide). Beef and dairy farming is particularly challenging in terms of climate mitigation with very few options for emissions reduction [17]. Hence, it is very difficult to reconcile growth in beef and dairy farming with a low GHG emissions economy. This results in a considerable challenge for Ireland to meet deep emissions reduction targets.

Ireland has not established a firm mandatory target for the year 2050, but does have ambitious and legally binding targets for GHG emissions reduction targets for the year 2020 (this is dealt with in detail in a separate paper [18]). Under Directive 2009/29/EC approximately half of GHG emissions are due to large point source emitters (within part of industry, power generation and transformation) and are regulated under the European Emissions Trading Scheme (ETS). The collective target for all participants in the EU ETS is a 21% reduction in GHG emissions relative to 2005 levels[†] by 2020. Under the EU Effort Sharing Decision 2009/406/EC for the remaining half of greenhouse gas emissions (including agriculture), i.e. non-ETS emissions, the target for Ireland is to achieve a 20% reduction relative to 2005 levels.

[†] For the period beyond 2020, Directive 2009/29/EC assumes ETS emissions reduce by 1.74% per annum

Renewable energies are one of the key drivers for significant reductions in GHG levels. Bioenergy in the form of bioliquids, biogas and solid biomass may have a major role to play and represent one of the major options for substituting fossil fuels in the energy mix. However there are a number of environmental concerns associated with bioenergy centering on potential ecosystem damage, especially in the developing countries, and the level of climate change benefits of some bioenergies, particularly first generation biofuels [19-22]. Arising from these concerns and those linked to impacts for food prices, the EU Renewable Energy Directive (Directive 2009/28/EC) [6] establishes that biofuels must meet certain "sustainability criteria" in order for them to be counted towards national biofuels targets. The main criteria are:

- From January 2017, the greenhouse gas emissions saving from the use of biofuels and bioliquids compared with the fossil fuels they displace shall be at least 50%. From 2018 that saving shall be at least 60%;
- Biofuels from peatlands and land with high biodiversity value or high carbon stock may not be used;
- Impact of biofuel policy on social sustainability, food prices and other development issues is to be assessed. Separate studies for Ireland [19, 23, 24] and UK [25] show these "sustainability criteria" beyond 2017 may affect the availability of bioenergy (especially biodiesel) from international trade limiting *de facto* the capacity of single countries of achieving emissions reduction targets.

The work presented in this paper assesses the role of bioenergy in Ireland in the context of achieving challenging GHG emission mitigation policies by 2050. It investigates a number of key technical and non-technical issues namely; how bioenergy can contribute to GHG emissions reduction targets for Ireland; how bioenergy will impact on Ireland's energy import dependency and how bioenergy will compete on land usage with agri-food sector. Moreover this work scrutinizes how limited availability of bioenergy imports impacts on the energy system attempts to achieve deep GHG emissions reductions. This analysis is carried using the Irish TIMES model which is a bottom-up technology rich energy systems model for Ireland (details in the Methodology section).

METHODOLOGY

The analysis in this report derives from scenario analysis using the Irish TIMES energy systems model [26]. The Irish TIMES model provides a range of energy system configurations for Ireland that each delivers projected energy service demand requirements optimised to least cost and subject to a range of technical and policy constraints for the period out to 2050. It provides a means of testing energy policy choices and scenarios, and assessing the implications:

- For the Irish economy (technology choices, prices, output, etc.);
- For Ireland's energy mix and energy dependence;
- For the environment, with a particular focus on greenhouse gas emissions.

It is used both to examine baseline projections, and to assess the implications of emerging technologies and of mobilising alternative policy choices such as meeting renewable energy targets and carbon mitigation strategies.

The Irish TIMES model was developed with TIMES (The Integrated Markal-Efom System) energy systems modelling tool; developed and supported by the Energy Technology Systems Analysis Program (ETSAP), an Implementing Agreement of the International Energy Agency (IEA) ‡. TIMES is a bottom-up model generator for local, national or multi-regional energy systems, which combines two different, but

‡See http://iea-etsap.org/ for more details

complementary, systematic approaches to modelling energy: a technical engineering approach and an economic approach [27]. TIMES computes a dynamic inter-temporal partial equilibrium on integrated energy markets. The objective function to maximize is the total surplus. This is equivalent to minimizing the total discounted energy system cost while respecting environmental and many technical constraints. This cost includes investment costs, operation and maintenance costs, plus the costs of imported fuels, minus the incomes of exported fuels, minus the residual value of technologies at the end of the horizon. The full technical documentation of the TIMES model is available in Loulou *et al.* [28]. A number of studies involving TIMES (and its predecessor MARKAL) models may be found in [29, 30].

The Irish TIMES model was originally extracted from the Pan European TIMES (PET) model and then updated with improved data based on much extensive local knowledge. The Irish energy system is characterized and modelled in terms of its supply sectors, its power generation sector, and its demand sectors. Extensive description and details on modelling structure and approach may be found in [18, 26, 31].

Model sets and assumptions

The Irish TIMES model used in this analysis has a time horizon of 45 years that ranges from 2005, the base year, to 2050, with a time resolution of four seasons with day-night time resolution, the latter comprising day, night and peak time-slices [26]. Energy demands are driven by a macroeconomic scenario, which is based on the ESRI HERMES macroeconomic model of the economy [32], with key drivers extended to the period 2050. On the supply side, fossil fuel prices are based on IEA's current policy scenario in World Energy Outlook 2012 report [33]. Given the importance of renewable energy for the achievement of mitigation targets, Ireland's energy potentials and costs are based on the most recently available data. The domestic bioenergy resources are represented by 12 different commodities. The total resource capacity limit for domestic bioenergy – considering both available and technical potential – has been set at 2,887 ktoe for the year 2030 and at 3,805 ktoe by 2050, based on the estimates from [12, 19, 23, 34, 35]. The potential for each individual commodity is shown in Table 2. The upper capacity limit for other renewable resources such as onshore and offshore wind energy, ocean, hydro, solar and geothermal energy are summarized in [31]. The use of geothermal energy in Ireland is limited only to small installations in the residential and services sector mostly for space and water heating purposes. Because solar and geothermal energy contribute marginally to scenarios outputs, no maximum potentials have been provided in the model.

The cost assumptions for domestic bioenergy commodities are based on [36] for biogas from grass, [37] for forestry, [38] for willow and miscanthus crops and delivery costs, and [23] for wheat crops, Oil Seed Rape (OSR) and Recycled Vegetable Oil (RVO). For the remaining commodities, the cost assumptions used in the PET model within the RES2020 project [39] were used. Cost estimates for bioenergy imports are based on [23] international trends. Details are summarized in Table 3. Cost assumptions for bulk renewable energy technologies are based on [40], [41] (for wind energy) and [42] (for solar).

Based on work undertaken by Ireland's transmission system operator EirGrid [43], the level of intermittent (non-dispatchable) renewable generation – namely wind, solar and ocean energy – is limited here to a maximum share of 70% of electricity generation within each timeslice and to 50% at annual level to account for operational issues associated with such high levels of variable generation in the power system. Regarding policies, investment subsidies and feed-in-tariffs for renewables based on policies currently in practice are assumed here to continue until 2030 and no trading of green certificates is assumed. The installation of new coal power plant capacities are limited to the replacement of current capacity levels, while for wind a maximum installation rate is set at 750 MW per year. Additional information regarding the main input assumptions may be found online at http://www.ucc.ie/en/energypolicy/irishtimes/.

In this analysis we do not model non energy-related emissions associated with agriculture but rather take projections from other sources and use them to exogenously establish the target for the energy system. Set against this backdrop, this paper makes a simple assumption regarding GHG emissions in agriculture, namely that agriculture emissions in 2020-2050 are the same as current national projections (+1% relative to 1990) [44] for 2020. This anticipates growth in agricultural activity in conjunction with the implementation of some level of mitigation.

Table 2. Bioenergy potential in the Irish TIMES model

Commodity	2010	2020	2030	2040	2050	Unit
Agricultural residues-dry	153	188	188	188	188	[ktoe]
Maize/wheat	0	42	45	45	45	[ktoe]
Miscanthus crop (Total)	6	36	160	285	353	[ktoe]
Miscanthus crop - RSV 1	6	36	89	89	89	[ktoe]
Miscanthus crop - RSV 2	0	0	22	22	22	[ktoe]
Miscanthus crop - RSV 3	0	0	0.2	0.2	0.2	[ktoe]
Miscanthus crop - RSV 4	0	0	37	37	37	[ktoe]
Miscanthus crop - RSV 5	0	0	7	7	7	[ktoe]
Miscanthus crop - RSV 6	0	0	3	22	22	[ktoe]
Miscanthus crop - RSV 7	0	0	0	106	174	[ktoe]
Willow crop (Total)	6	33	143	255	316	[ktoe]
Willow crop - RSV 1	6	33	79	79	79	[ktoe]
Willow crop - RSV 2	0	0	8	8	8	[ktoe]
Willow crop - RSV 3	0	0	12	12	12	[ktoe]
Willow crop - RSV 4	0	0	20	20	20	[ktoe]
Willow crop - RSV 5	0	0	25	40	40	[ktoe]
Willow crop - RSV 6	0	0	0	12	12	[ktoe]
Willow crop - RSV 7	0	0	0	85	146	[ktoe]
Forestry residues	122	176	212	269	326	[ktoe]
Biogas from landfill and other	57	57	57	57	57	[ktoe]
Biogas from Grass	0	744	1,136	1,136	1,136	[ktoe]
Municipal waste - BMSW	142	543	706	869	1,031	[ktoe]
Recycled vegetable oil	0	1	2	2	2	[ktoe]
Oil seed rape/algae	2	30	41	95	133	[ktoe]
Agricultural residues - wet	67	78	79	79	79	[ktoe]
Wood processing residues	75	92	117	115	137	[ktoe]
	630	2,021	2,887	3,395	3,805	[ktoe]

Table 3. Bioenergy cost assumption ($€_{2000}$/GJ)

Commodity	2010	2020	2030	2040	2050	Unit
Agricultural residues-dry	4.6	5.2	5.2	5.2	5.2	[€/GJ]
Maize/wheat	17.7	17.7	17.7	18.7	19.8	[€/GJ]
Miscanthus crop - RSV 1	2.8	4.4	4.8	5.1	5.4	[€/GJ]
Miscanthus crop - RSV 2	3.0	4.8	5.3	5.6	5.9	[€/GJ]
Miscanthus crop - RSV 3	3.3	5.3	5.8	6.1	6.4	[€/GJ]
Miscanthus crop - RSV 4	3.6	5.7	6.3	6.6	7.0	[€/GJ]
Miscanthus crop - RSV 5	3.9	6.1	6.7	7.1	7.5	[€/GJ]
Miscanthus crop - RSV 6	4.1	6.6	7.2	7.6	8.1	[€/GJ]
Miscanthus crop - RSV 7	4.4	7.0	7.7	8.1	8.6	[€/GJ]
Willow crop - RSV 1	4.3	6.9	7.6	8.0	8.4	[€/GJ]
Willow crop - RSV 2	4.8	7.6	8.3	8.8	9.3	[€/GJ]
Willow crop - RSV 3	5.2	8.3	9.1	9.6	10.1	[€/GJ]
Willow crop - RSV 4	5.6	8.9	9.8	10.4	11.0	[€/GJ]
Willow crop - RSV 5	6.0	9.6	10.6	11.2	11.8	[€/GJ]
Willow crop - RSV 6	6.5	10.3	11.4	12.0	12.7	[€/GJ]
Willow crop - RSV 7	6.9	11.0	12.1	12.8	13.5	[€/GJ]
Forestry residues	6.8	6.8	6.8	6.8	6.8	[€/GJ]
Biogas from landfill and other	3.3	4.7	5.1	5.4	5.6	[€/GJ]
Biogas from grass	6.6	6.6	6.6	6.9	7.2	[€/GJ]
Municipal waste - BMSW	0.4	0.2	0.2	0.2	0.2	[€/GJ]
Recycled vegetable oil	5.5	5.5	5.5	5.8	6.1	[€/GJ]
Oil seed rape/algae	23.1	23.1	23.1	23.1	23.1	[€/GJ]
Agricultural residues - wet	0.0	0.0	0.0	0.0	0.0	[€/GJ]
Wood processing residues	1.9	1.9	1.9	1.9	1.9	[€/GJ]
Bio ethanol - RSV 1	19.0	18.0	16.2	16.2	16.2	[€/GJ]
Bio ethanol - RSV 2	19.0	19.5	19.5	20.6	21.8	[€/GJ]
Bio ethanol - RSV 3	19.0	21.1	24.0	25.4	26.8	[€/GJ]
Bio ethanol - RSV 4	19.0	23.2	29.4	31.0	32.7	[€/GJ]
Biodiesel - RSV 1	26.6	30.6	28.9	28.9	28.9	[€/GJ]
Biodiesel - RSV 2	26.6	33.0	34.1	36.0	38.0	[€/GJ]
Biodiesel - RSV 3	26.6	35.3	40.3	42.6	45.0	[€/GJ]
Biodiesel - RSV 4	26.6	38.6	48.7	51.4	54.3	[€/GJ]
Wood pellets - RSV 1	11.0	6.7	5.4	5.4	5.4	[€/GJ]
Wood pellets - RSV 2	11.0	7.1	6.2	6.2	6.2	[€/GJ]
Wood pellets - RSV 3	11.0	7.9	6.9	6.9	6.9	[€/GJ]
Wood pellets - RSV 4	11.0	8.5	7.9	7.9	7.9	[€/GJ]
Bio rape seed	31.1	33.3	35.6	37.8	40.0	[€/GJ]
Wood chip - RSV 1	5.4	3.3	2.7	2.7	2.7	[€/GJ]
Wood chip - RSV 2	5.4	3.5	3.0	3.0	3.0	[€/GJ]
Wood chip - RSV 3	5.4	3.9	3.4	3.4	3.4	[€/GJ]
Wood chip - RSV 4	5.4	4.2	3.9	3.9	3.9	[€/GJ]

Scenarios

In this paper results for four distinct scenarios are presented to explore the role of bioenergy in Ireland's low carbon future. The main scenarios assumptions are listed below:

- Business as Usual (BaU) scenario: it delivers energy system demands at least cost in the absence of emissions reduction targets and efficiency improvements. It is used as a reference case (counterfactual) against which to compare three the distinct mitigation scenarios;

- CO2-80 scenario: the energy system is required to achieve at least an 80% CO_2 emissions reduction below 1990 levels by 2050 (-85.7% relative to 2005). The pathway includes interim targets in line[§] with the EU 2020 climate energy package [4, 5], i.e. 20% CO_2 emissions reduction by 2020 relative to 2005 levels. Agriculture GHG emissions are implicitly assumed to grow by 4% in the period 2005-2020 [44], while over the period 2020-2050 are assumed constant;

- CO2-80 SC scenario: it delivers the same emissions reduction pathway than CO2-80 scenario, but it simulates how shortages on imported bioenergy commodities consequent with the introduction of the Sustainability Criteria (SC) of the EU Renewable Energy Directive may affect the energy system choices. To simulate the maximum levels of available imported bioenergy, which meet SC requirements, we refer to analysis in Clancy *et al.* [23]. Assuming a global context of high bioenergy demand driven by the introduction of mitigation targets in several countries, the "Medium supply/High demand" scenario has been used as main reference for the period 2010-2030, as shown in Table 4[**].

Table 4. Imported bioenergy potential in CO2-80 SC scenario

Description	2010	2020	2030	2040	2050	Unit
Bio ethanol	781.3	409.4	1,404.1	1,460.8	1,519.9	[ktoe]
Biodiesel	101.5	0.0	109.9	114.4	119.0	[ktoe]
Wood pellets	22.9	0.0	427.1	444.4	462.4	[ktoe]
Wood chip	7.6	0.0	142.4	148.1	154.1	[ktoe]

- CO2-80 DR scenario: it delivers the same emissions reduction pathway than CO2-80 scenario, but it simulates an energy scenario where, given the growing concerns over sustainability and impacts in terms of Direct and Indirect Land Use Change (DLUC and ILUC) of most of the imported bioenergy crops, the mitigation targets may be achieved only by mean of Domestic Resources (DR), meaning that no bioenergy imports are allowed beyond 2020.

The main scenarios assumptions are summarized in Table 5.

RESULTS

This section provides a range of energy system configurations for Ireland that each deliver projected energy service demand requirements optimised to least cost and subject to different policy constraints for the time period out to 2050. This provides a means of testing energy policy choices and scenarios and assesses the implications for the Irish economy and energy system. This results section is structured as follows. Firstly

---

[§] although not with the ETS / non-ETS split
[**] Beyond 2030 we assumed a 2% increase every 5 years

pathways are presented and discussed. Secondly the BaU scenario is compared against the CO2-80 scenario focusing in particular on the role of bioenergy. This section further discusses implications for the energy system of reduced availability of sustainable bioenergy for international trade, assessing how this results in terms of capacity of delivering deep emissions reductions. This is followed by a discussion on how these future low carbon economies may result on Irelands import dependency and land usage. Lastly it discusses economic impacts of energy futures in term of CO_2 marginal abatement costs.

Table 5. Summary of scenario assumptions

Scenario	Mitigation target			Bioenergy imports		
	2020	2030	2050	2020	2030	2050
BaU	No	No	No	Yes	Yes	Yes
CO2-80	-20.4% GHG (-30.4% CO_2) rel. 2005	-20% GHG (-32% CO_2) rel. 1990	-52.4% GHG (-80% CO_2) rel. 1990	Yes	Yes	Yes
CO2-80 SC	-20.4% GHG (-30.4% CO_2) rel. 2005	-20% GHG (-32% CO_2) rel. 1990	-52.4% GHG (-80% CO_2) rel. 1990	Limited	Limited	Limited
CO2-80 DR	-20.4% GHG (-30.4% CO_2) rel. 2005	-20% GHG (-32% CO_2) rel. 1990	-52.4% GHG (-80% CO_2) rel. 1990	No	No	No

The total GHG emissions pathways for the four policy scenarios are shown in Figure 3.

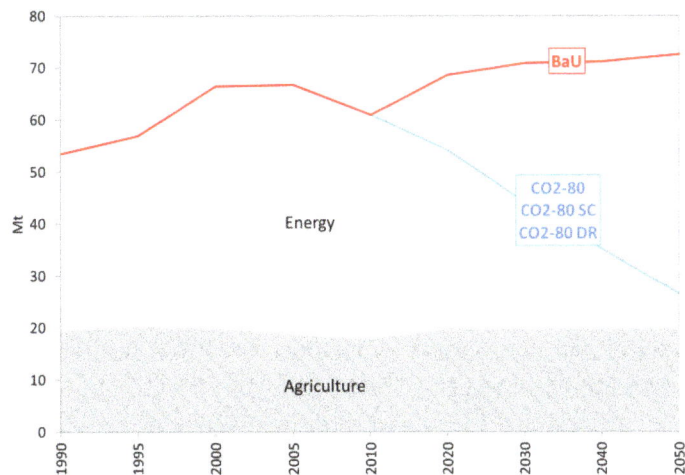

Figure 3. Total GHG emissions trajectories by scenario (Mt $CO_{2,eq}$)

The role of bioenergy in Ireland's low carbon future

This section firstly presents the CO_2 emissions for the resultant energy systems from the BaU and 80% CO_2 emissions reduction scenario (CO2-80) for the period to 2050. The results show radically different futures. In the absence of emissions mitigation (see Figure 4), the BaU scenario shows the energy system emissions at approximately 53 Mt

CO_2 in 2050, representing a growth of 24% relative to 2010 (or 55% growth relative to 1990). By contrast, an 80% CO_2 reduction target means effectively reducing by 87% the projected BaU emissions. In the CO2-80 scenario the greatest reduction in emissions relative to 2010 is in the transport sector (from 19.6 Mt to 1.7 Mt) followed then by electricity generation (from 14.3 Mt to 0.9 Mt).

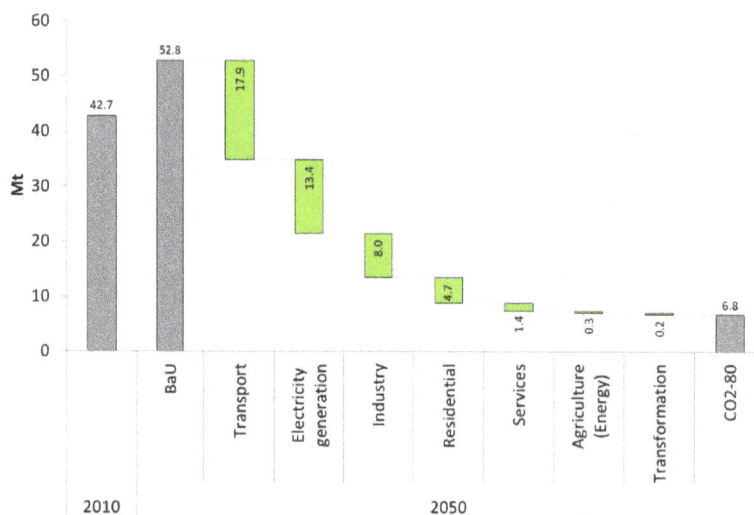

Figure 4. Incremental change in CO_2 emission required by each sector in CO2-80 relative BaU scenario and 2010 (Mt)

The evolution of Total Final Consumption (TFC) of energy by sector in BaU and CO2-80 scenarios is presented in Figure 5. Changes in final energy consumption are driven by economic activity (which affects energy service demands), the type of end use energy (including electricity) and the efficiencies of end-use technologies, in addition to consumer response to changing energy prices and to policy measures. There is currently no feedback between the Irish TIMES scenario results and the economy and hence in all scenarios, economic growth (measured in terms of GDP) follows the same trend, growing by 1.9% per annum on average over the period 2010-2050. TFC grows by 0.9% p.a. in the BaU scenario and remain stable (+0.002% p.a.) in the CO2-80 scenario, illustrating the increased decoupling between economic growth and emissions growth.

Table 6 and Figure 5 summarize the primary energy requirements in these alternative energy futures. The projected primary energy consumption in the BaU suggests future trends very similar to current, i.e. substantial reliance on oil and gas with a small share for renewables. The CO2-80 scenario shows a drop in reliance on oil from 2030, coupled with a renewables (wind and bioenergy) expansion. By 2050 liquid biofuels and biogas are extensively used in transport (51%-55% of transport TFC), while biomass is largely used in industry (63% of industry TFC) and buildings (26% of buildings TFC). Coal has all but disappeared from the domestic energy system except for use in industry in combination with CCS technology. Wind energy and natural gas in combination with CCS technology provide an impetus for the electrification of private cars and rail in the transport sector.

Focussing on renewable energy and in particular on bioenergy Figure 6 details the modal results for renewable heat (RES-H), transport (RES-T) and electricity (RES-E) from the energy system cost optimal analysis for the BaU scenario and the CO2-80 scenario. The coloured areas represent bioenergy commodities, while grey pattern bars represent other renewables. In the BaU scenario bioenergy consumptions are dominated by biomass, with 642 ktoe in RES-H while bioethanol dominates in the transport sector (93 ktoe). In the RES-E 46 ktoe are delivered by biogas generation. In the CO2-80 scenario growth of bioenergy consumptions are mainly delivered in the heating sector, where biomass consumption results 2,500 ktoe by 2050 and transport; and transport, which shows an steep increase of biofuel consumption, delivered by bioethanol (33.7%), biogas (28.2%) and biodiesel (28.0%). No bioenergy is hence consumed in the electricity generation sector, dominated by wind generation (94.1% of renewable electricity). In this scenario (CO2-80), renewables account for 61.8% of Gross Fuel Consumption (GFC)[††] by 2050, in which biofuels deliver for 81.4% of transport GFC[‡‡] and 61.4% of thermal GFC.

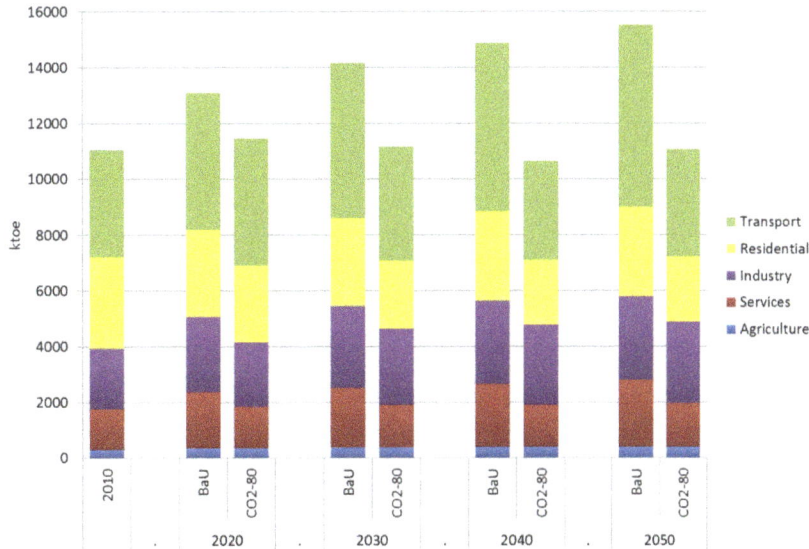

Figure 5. Final energy demand by sector in REF and CO2-80 (ktoe)

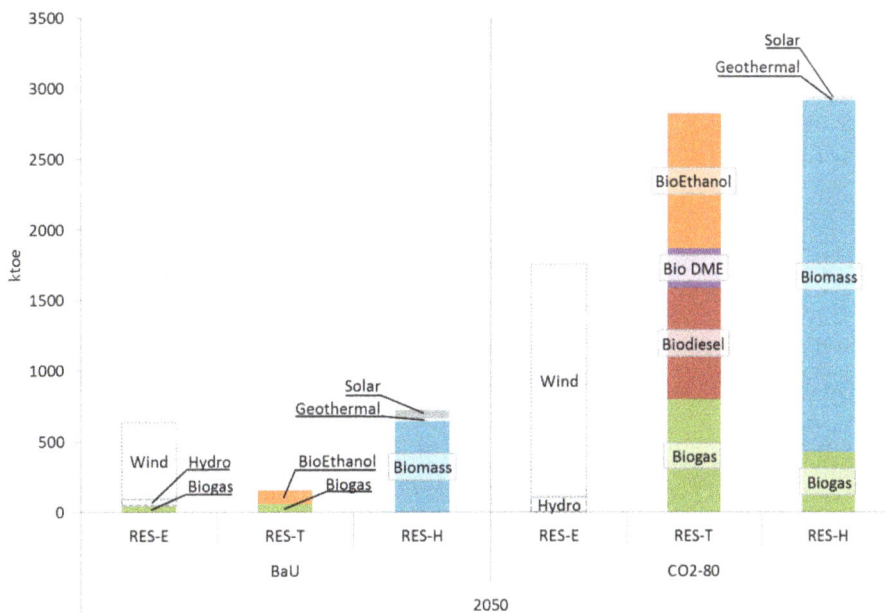

Figure 6. Bioenergy consumption by mode in BaU and CO2-80 (ktoe)

[††] Instructions from Article 5 of Directive 2009/28/EC have been used as reference for the calculation
[‡‡] Excluding international aviation

Table 6. Primary energy trends for BaU, CO2-80 (ktoe)

	2010	BaU 2020	2030	2050	CO2-80 2020	2030	2050
Fossil Fuels (total)	14,436	17,211	18,384	19,406	13,507	11,225	7,859
Coal and peat	2,031	2,325	2,008	1,436	715	456	506
Oil (incl. int. aviation)	7,713	9,807	10,499	11,395	7,917	6,464	3,055
Oil (excl. int. aviation)	6,939	8,484	9,082	9,871	6,594	5,047	1,530
Natural gas	4,692	5,078	5,878	6,575	4,875	4,305	4,298
Renewables (total)	761	1,347	1,530	1,863	1,466	4,525	11,286
Hydro	52	45	45	43	88	93	104
Wind	242	545	545	545	662	1,334	1,651
Biomass	211	552	641	828	415	1,489	3,071
(of which imported)	11	3	38	117	58	877	1,873
Bioliquids	93	73	120	96	102	352	1,751
(of which imported)	71	71	77	93	71	309	1,615
Biogas	58	57	57	57	57	57	1,193
Other renewables	24	0	6	82	12	13	28
Electricity imports (net)	40	0	55	119	170	170	170
Total	15,238	18,558	19,969	21,388	15,143	15,921	19,315

How sustainable is the low carbon future?

This section discusses the sustainability of low carbon future pathways, in particular how the ability of the energy system of delivering deep reductions in emissions levels given the sustainability implications of bioenergy imports. The CO2-80 results are compared with results from the CO2-80 SC scenario (which limits imported bioenergy commodities due to the Sustainability Criteria (SC) of the EU Renewable Energy Directive) and the CO2-80 DR scenario (in which mitigation targets may be achieved only by mean of Domestic Resources (DR)).

Figure 7, which compares bioenergy and other renewables consumption by sector[§§] for CO2-80, CO2-80 SC and CO2-80 DR, shows that restrictions in biofuels and biomass imports have only limited impact on the short term (2020) but may have a larger impact on over the longer term. Reductions in bioenergy levels are only partially replaced with domestic bioenergy resources and other renewable sources (mostly from the power sector).

Results for the CO2-80 SC scenario indicate that since 2030 bioenergy consumption reduces in all end-use sectors (by about 6% in 2030 and by 19% in 2050 relative to the CO2-80 scenario) while renewable electricity grows (+36% by 2050 relative to CO2-80). Figure 8 shows that in the transport sector the drop in biodiesel imports are only partially balanced by higher domestic biogas production (from grass) (+21%) and increased imports of ethanol (+32%). With respect to heating, electricity displaces biomass and biogas (-20% and -38%) in the heating sectors.

The CO2-80 DR shows a similar pattern, but with steeper reduction trends in bioenergy consumption. By 2030 the reduction in bioenergy consumption is 36% lower relative to the unconstrained case (CO2-80) and passes to 53% in 2050. The heating sectors moves further from bioenergy (-42% in 2030 and -45% in 2050) to electricity (+2.5% in 2030 and +76% in 2050) which shows increased levels of renewable

---

[§§] In figure the electricity generation sector is classed as ELC, residential sector as RSD, services as SRV, agriculture as AGR, and transport as TRA

generation (+2.4% in 2030 and +70% in 2050 from onshore and offshore wind, solar and some ocean energy). The transport sector (freight and public transport) from 2030 transitions from bioliquids to biogas, while in 2050 about 40% of freight fleet consumes hydrogen (from gasification of coal with CCS).

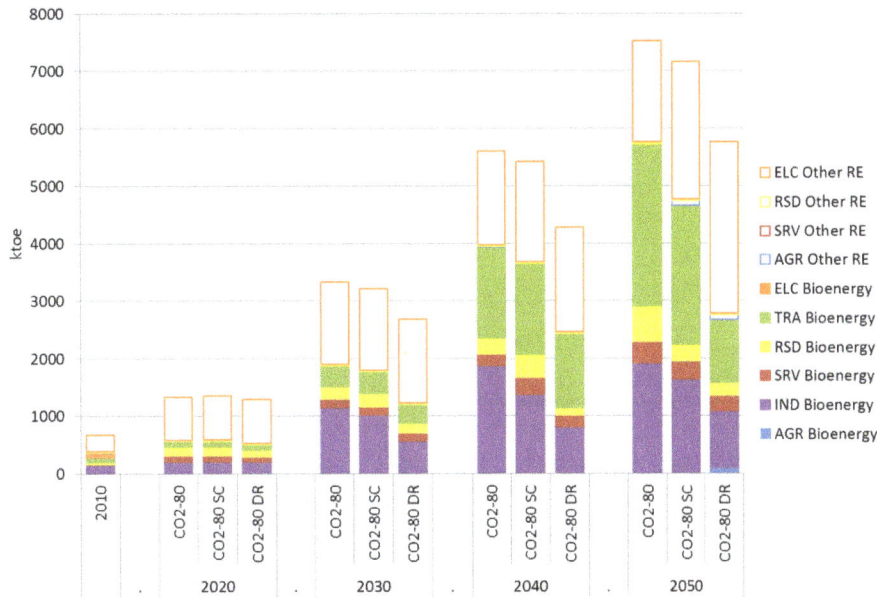

Figure 7. Bioenergy and other renewables consumption by sector in CO2-80, CO2-80 SC and CO2-80 DR (ktoe)

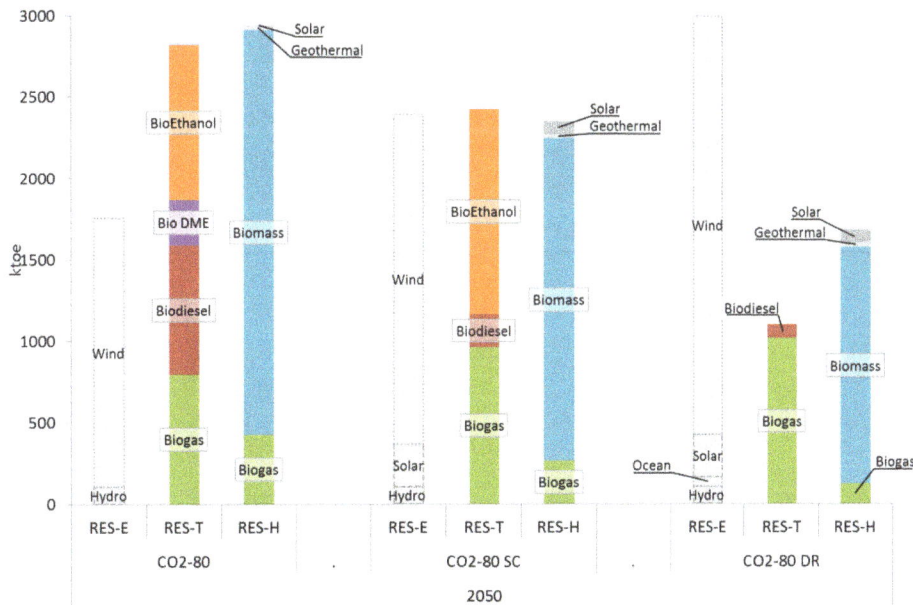

Figure 8. Bioenergy consumption by mode in CO2-80, CO2-80 SC and CO2-80 DR (ktoe)

The results (shown in Table 7) moreover indicate that drops on bioenergy imports cause reductions of the renewable shares (measured as share of total gross energy consumption)[***], which are driven by reduced bioenergy consumptions in transport and

---

[***] Instructions from Article 5 of Directive 2009/28/EC have been used as reference for the calculation.

heating sectors (where biomass and biofuels are the dominant renewable sources). These limitations do not influence the share in the electricity generation sector, where it indeed grows. By contrast, the reduced bioenergy availability forces the model to adopt deeper efficiency measures causing an increase in end-use efficiency in transport, residential and services sectors.

Table 7. Renewable share and energy efficiency for CO2-80, CO2-80 SC and CO2-80 DR

	CO2-80		CO2-80 SC		CO2-80 DR	
	2030	2050	2030	2050	2030	2050
Renewable share	25.4%	54.8%	24.7%	52.2%	21.1%	43.6%
of which RES-H	11.7%	21.2%	10.9%	17.3%	7.0%	12.7%
of which RES-E	11.0%	13.0%	10.9%	17.6%	11.4%	22.6%
of which RES-T	2.9%	20.9%	3.0%	17.8%	2.6%	8.3%
Energy savings	-21.3%	-28.7%	-22.0%	-29.2%	-23.3%	-32.9%

As third consequence, the results highlight (Table 8) an increase in electricity importance for the end-use sectors. By 2050 electricity grows by 35% (CO2-80 SC) and 67% (CO2-80 DR) respectively relative to the CO2-80 case; and become the most important energy vector for meeting heat demand. The electricity generation fuel mix to meet this increased electricity demand is summarized in Figure 9.

Table 8. Share of energy use in end-use sectors for CO2-80, CO2-80 SC and CO2-80 DR

	CO2-80		CO2-80 SC		CO2-80 DR	
	2030	2050	2030	2050	2030	2050
Fossil fuels/TFC	64.3%	30.5%	65.0%	28.6%	68.3%	33.1%
Renewables/TFC	14.9%	44.5%	14.2%	37.4%	9.9%	23.1%
Electricity/TFC	20.8%	25.0%	20.8%	34.0%	21.8%	43.8%
of thermal TFC	33.3%	35.5%	33.2%	52.2%	35.1%	65.9%
of transport TFC	4.9%	11.0%	5.2%	11.0%	5.3%	12.3%

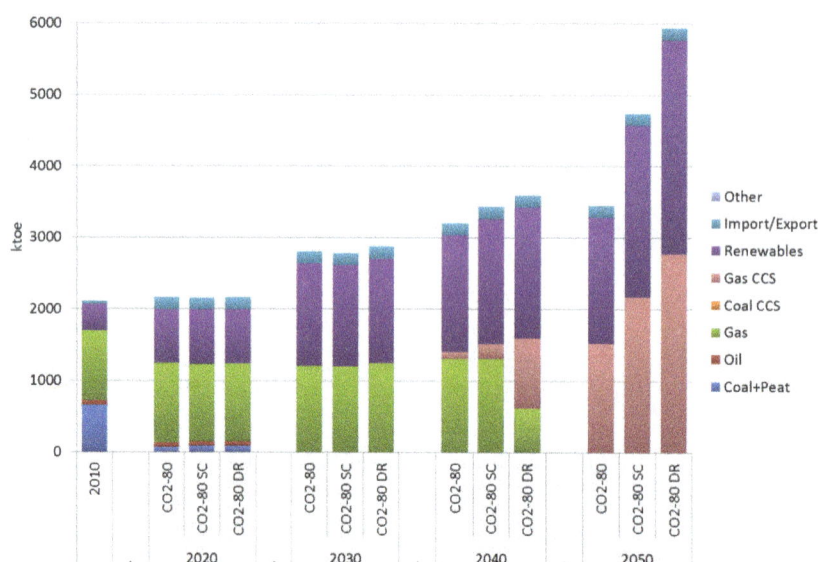

Figure 9. Electricity generation by fuel in CO2-80, CO2-80 SC and CO2-80 DR

Bioenergy and import dependency

Table 9 highlights the implications of these different mitigation scenarios on another key policy issue, energy security. The analysis here is limited to import dependency, which is a crude and limited metric by which to assess energy security. More details on implications for Ireland's energy security are assessed in a separate analysis [45]. Focussing first on primary energy import dependency, the results show that the import dependency in the business as usual scenario grows to approximately 93% in 2050, while across the mitigation scenarios reducing trends are shown. Bioenergy contributes in this reduction resulting in all scenarios with lower import dependency indices compared to overall primary energy levels.

Table 9. Primary energy and bioenergy import dependency

	Scenario	2010	2020	2030	2040	2050
Primary energy	BaU	86.0%	86.8%	87.9%	91.4%	92.9%
	CO2-80	86.0%	85.9%	81.3%	78.3%	72.4%
	CO2-80 SC	86.0%	85.1%	79.6%	71.4%	67.5%
	CO2-80 DR	86.0%	84.9%	76.9%	68.9%	65.0%
Bioenergy	BaU	32.6%	10.9%	14.1%	16.6%	21.5%
	CO2-80	32.6%	22.4%	62.5%	68.6%	58.0%
	CO2-80 SC	32.6%	13.7%	47.3%	41.9%	41.3%
	CO2-80 DR	32.6%	0.0%	0.0%	0.0%	0.0%

Bioenergy and land usage

The potential growth of bioenergy raises a number of concerns relating to land depletion and implications with one of Ireland's most important economic sectors: the agri-food sector. These concerns have also been highlighted recently in [24] which shows that the EU Agricultural Policy (Cross Compliance) [46, 47] does not accept that pasture (currently 4 Mha in Ireland) can be ploughed to generate arable land for biofuel production. Ireland is not self-sufficient in grains [19] and as such there would be intense competition for a grain ethanol industry with the likelihood that ethanol production in Ireland would be based on imported grains or at least necessitate import of more grain [24].

This section therefore presents a first attempt on quantifying this impact, presenting modelling results, not only in terms of energy flows or emissions, but also in terms of land consumption. The conversion factors of each individual commodity (Table 10) are drawn from [24, 48, 49]. Crop rotation levels determine the ratio between required and contracted land.

Table 10. Bioenergy conversion factors

Commodity	Conversion factor [ha/ktoe]	Rotation	Reference
Willow	253.0	1 in 2	[48]
Miscanthus	268.3	1 in 1	[48]
Rape seed biodiesel	910.2	1 in 5	[49]
Palm oil biodiesel	348.9	1 in 1	[49]
Wheat ethanol	634.4	2 in 3	[49]
Optimized wheat ethanol	498.4	2 in 3	[49]
Grass biomethane	263.4	1 in 1	[24]

Table 11 summarizes energy crops (including grass) consumptions in the different scenarios converted into land units, namely hectares. Regarding imported commodities, the model does not distinguish between different import locations nor different feedstock crops and hence the following assumptions were made to complete this analysis:

- Imported bioethanol is assumed to originate from optimized wheat crops;
- Biodiesel originates from palm oil;
- Woody biomass originates from miscanthus crops.

Given the total of Ireland's agriculture land is 4.3 Mha [49], the required land for domestic energy crops in 2030 ranges from 1.4% (in BaU) to 4% (in CO2-80 DR) and by 2050 between 0.7% (in BaU) and 11.9% (in CO2-80 DR) of total agriculture land. Given crop rotation this translates into values shown in Table 11. Equally bioenergy imports by 2050 require the equivalent of 1.8% (BaU) to 28.3% (CO2-80) of current agricultural land.

Currently tillage accounts only for about 0.4 Mha, while the remaining 3.9 Mha are under pasture grassland. Mitigation scenarios therefore indicate that by 2050 to produce methane from grass would require the equivalent of 8% of current grassland area. Energy crops in total (willow, miscanthus, wheat and rapeseed) would require an equivalent between 64% (CO2-80) and 73% (CO2-80 DR) of today's arable land contracted by 2030 and between 79% and 113% by 2050.

However research in [50] has highlighted that practices such as increasing nitrogen (N) fertiliser input (to the limit permitted by the EU Nitrates Directive) combined with increasing the grazed grass utilisation rate may in future significantly increase the grass resource available in excess of livestock requirements (from 1.7 million t of dry matter (DM) to 12.2 million t DM/annum), limiting the competition with traditional dairy, beef and lamb production systems (reduction in required land for energy uses) and providing an alternative enterprise and income to farmers. This will potentially reduce land use competition by making more land readily available for grass as a feedstock for biomethane production.

Table 11. Land required (contracted) for domestic and imported energy crops in 2030 and 2050

Unit: [kha]		BaU		CO2-80		CO2-80 SC		CO2-80 DR	
		2030	2050	2030	2050	2030	2050	2030	2050
Domestic	Willow	0	0	40	44	50	160	73	160
	Miscanthus	30	30	43	95	43	95	43	95
	Grass biomethane	0	0	0	299	0	299	63	299
	Wheat ethanol	0	0	0	0	0	0	0	8
	Rape seed biodiesel	148	0	148	148	148	148	148	148
	TOTAL	178	30	231	586	241	702	326	710
Imported	Wheat ethanol	58	70	213	677	222	894	0	0
	Palm oil biodiesel	0	0	8	247	8	42	0	0
	Wood chip[†††]	10	31	229	229	36	40	0	0
	Wood pellets[‡‡‡]	0	0	0	260	102	121	0	0
	TOTAL	68	100	450	1,413	368	1,096	0	0
[%] of agri land	Domestic	4.2	0.7	5.4	13.7	5.7	16.5	7.6	16.7
	Imported	1.6	2.4	10.7	33.6	8.8	26.1	0	0

[†††] from Miscanthus
[‡‡‡] from Miscanthus

Marginal abatement cost of mitigation targets

One of the main insights that can be gained from the use of energy systems models such as TIMES is from quantifying the impact of different mitigation targets on marginal CO_2 abatement costs, which provide an indication of the costs of abating the last tonne of CO_2 and can be used as a proxy for indicating the level of carbon tax that may be required to reach a certain level of mitigation.

Table 12 summarises the marginal CO_2 abatement costs for the mitigation scenarios presented in this paper. The CO2-80 SC scenario indicates as early as 2030, higher CO_2 abatement prices due to insufficient availability of bioenergy resources. By 2050 this difference becomes steeper, illustrating how bioenergy imports influences the achievement of this challenging mitigation targets. Similarly the CO2-80 DR scenario shows that limitations in import options may forces the energy system to invest in expensive abatement technologies (e.g. hydrogen) which drives the marginal abatement costs at values even higher than the CO2-80 SC case.

Table 12. CO_2 marginal abatement cost ($€_{2010}$/tonne)

Scenario	2020	2030	2040	2050	
CO2-80	74	98	312	395	[€/tonne]
CO2-80 SC	74	110	380	1,389	[€/tonne]
CO2-80 DR	74	259	387	1,747	[€/tonne]

CONCLUSIONS

Transitioning to a low carbon economy to mitigate climate change represents globally one of the most challenging policy targets for the future years. The EU has set this ambition implementing policy targets for the year 2020 and aiming to achieve reductions between 80% and 95% relative to 1990 levels by 2050. The paper has a specific focus on bioenergy, which the results suggest are likely to be the most significant fuel source for the future low carbon economy. There are several concerns however regarding sustainability of these energy sources. The paper shown that application of sustainability criteria in international markets – for example as in the EU Renewable Energy Directive – may cause restrictions in bioenergy supply (mostly biodiesel), which can strongly influence the ability of Ireland energy system to deliver GHG emissions reductions. With constraints on imports, bioenergy contributions are significantly reduced, mainly within the transport sector, with consequent increases in electrification – based on gas CCS and renewables (wind, solar and also ocean) – end-use efficiency and hydrogen. Marginal CO_2 abatement costs rise sharply in accordance with the level of import restrictions.

This paper also sheds light on some of the implications for energy security. The energy import dependency in Ireland is anticipated to be reduced significantly in all the mitigation scenarios considered. Variable renewable energies – namely wind, solar and ocean – are the main drivers of this reduction, but also bioenergy positively contributes with at least 40% domestic consumption.

Finally the results point to the implications of bioenergy in terms of land usage. Domestically bioenergy passes from approximately 5,000 ha of land contracted in 2010, to about 710,000 ha by 2050 (in the CO2-80 DR scenario), equivalent to 17% of total agricultural land area. This may have serious implications for the food supply which should be addressed in future. Further research work is required to improve the integrated modelling of both the energy and agriculture systems in order to provide richer insights to the strategy between energy, food and climate mitigation.

NOMENCLATURE

Abbreviations

CO_2,eq	Carbon Dioxide Equivalent
CO_2	Carbon Dioxide
EC	European Commission
EEA	European Environment Agency
ELC	Electricity Generation Sector
EPA	Environmental Protection Agency
ETS	Emissions Trading Scheme
ETSAP	Energy Technology Systems Analysis Program
EU	European Union
GDP	Gross Domestic Product
GFC	Gross Final Energy Consumption
GHG	Greenhouse Gases
ha	Hectare
IEA	International Energy Agency
IND	Industry Sector
Non-ETS	Non-Emission Trading Sectors
RSD	Residential Sector
RES-E	Renewables in the Electricity Sector
RES-H	Renewables in the Heating Sector
RES-T	Renewables in the Transport Sector
SRV	Services Sector
TFC	Total Final Consumption
TRA	Transport Sector

REFERENCES

1. UNFCCC, Report of the Conference of the Parties on its Fifteenth Session, held in Copenhagen from 7 to 19 December 2009. Decisions adopted by the Conference of the Parties, 2009, https://unfccc.int/documentation/documents/advanced_search/items/6911.php?priref=600005735#beg, [Accessed: October-2013]

2. IPCC, Climate Change 2014: Mitigation of Climate Change. Working Group III Contribution to the Fifth Assessment Report of the IPCC2014, Cambridge, United Kingdom and New York, NY, USA, Cambridge University Press.

3. IEA, Redrawing the Energy-Climate Map, http://www.worldenergyoutlook.org/energyclimatemap/, [Accessed: August-2013]

4. EU, DIRECTIVE 2009/29/EC the European Parliament and of the Council of 23 April 2009 amending Directive 2003/87/EC so as to improve and extend the Greenhouse Gas Emission Allowance Trading Scheme of the Community, European Parliament and Council, Editor 2009, *Official Journal of the European Union*, pp 25.

5. EU, DECISION No 406/2009/EC of the European Parliament and of the Council of 23 April 2009 on the effort of Member States to reduce their Greenhouse Gas Emissions to meet the Community's Greenhouse Gas Emission reduction Commitments up to 2020, European Parliament and Council, Editor 2009, *Official Journal of the European Union*, pp 13.

6. EC, DIRECTIVE 2009/28/EC of the European Parliament and of the Council (EC) of 23 April 2009 on the Promotion of the use of Energy from Renewable Sources and

Amending and Subsequently Repealing Directives 2001/77/EC and 2003/30/EC, http://www.energy.eu/directives/pro-re.pdf

7. EU, DIRECTIVE 2006/32/EC of the European Parliament and of the Council of 5 April 2006 on Energy end-use Efficiency and Energy Services and repealing Council Directive 93/76/EEC, European Parliament and Council, Editor 2006, *Official Journal of the European Union*, pp 22.

8. EC, COM(2014) 15 final, A Policy Framework for Climate and Energy in the Period from 2020 to 2030, Communication from the Commission to the European Parliament, the Council, the European Economic and Social Committee and the Committee of the Regions, http://ec.europa.eu/clima/policies/2030/docs/com_2014_15_en.pdf

9. EC, COM/2011/112, A Roadmap for moving to a Competitive Low Carbon Economy in 2050, Communication from the Commission to the European Parliament, the Council, the European Economic and Social Committee and the Committee of the Regions, European Commission, Editor 2011, Brussels, Belgium, pp 16.

10. EC, COM(2011) 885 Final, Energy Roadmap 2050, Communication from the Commission to the European Parliament, the Council, the European Economic and Social Committee and the Committee of the Regions, http://eur-lex.europa.eu/LexUriServ/LexUriServ.do?uri=COM:2011:0885:FIN:EN:PDF

11. EC, COM(2011) 144 Final, Roadmap to a Single European Transport Area – Towards a Competitive and Resource Efficient Transport System, http://eur-lex.europa.eu/LexUriServ/LexUriServ.do?uri=COM:2011:0144:FIN:EN:PDF

12. Howley, M., et al., Energy in Ireland 1990-2011, 2012 Report, Report published by Sustainable Energy Authority of Ireland, Dublin, Ireland, 2012.

13. IEA, Oil Information, Paris, France, 2012.

14. IEA, Natural Gas Information, Paris, France, 2012.

15. EEA, Annual European Union Greenhouse Gas Inventory 1990-2011 and Inventory Report 2013, Submission to the UNFCCC Secretariat, http://www.eea.europa.eu/data-and-maps/data/data-viewers/greenhouse-gases-viewer

16. CSO, Database of Area Farmed in June by Region, Type of Land Use and Year, http://www.cso.ie/px/pxeirestat/Statire/SelectVarVal/Define.asp?MainTable=AQA05&TabStrip=Select&PLanguage=0&FF=1, http://dataservice.eea.europa.eu/PivotApp/pivot.aspx?pivotid=475, [Accessed: October-2013]

17. Schulte, R., et al., A Marginal Abatement Cost Curve for Irish Agriculture, Teagasc Submission to the National Climate Policy Development Consultation, ed. R. Schulte and T. Donnellan, Carlow, Ireland: Report Published by Teagasc, 2012, http://www.teagasc.ie/publications/view_publication.aspx?PublicationID=1186

18. Chiodi, A., et al., Modelling the Impacts of Challenging 2020 non-ETS GHG Emissions reduction Targets on Ireland's Energy System, *Energy Policy*, Vol. 62, No. 0, pp 1438-1452, 2013.

19. Smyth, B. M., et al., Can we meet Targets for Biofuels and Renewable Energy in Transport given the Constraints imposed by Policy in Agriculture and Energy?, *Journal of Cleaner Production*, Vol. 18, No. 16-17, pp 1671-1685, 2010.

20. Börjesson, P., Good or Bad Bioethanol from a Greenhouse Gas Perspective – What determines this?, *Applied Energy*, Vol. 86, No. 5, pp 589-594, 2009.

21. Escobar, J. C., et al., Biofuels: Environment, Technology and Food Security, *Renewable and Sustainable Energy Reviews*, Vol. 13, No. 6-7, pp 1275-1287, 2009.

22. Thamsiriroj, T. and Murphy, J. D., Is it Better to import Palm Oil from Thailand to produce Biodiesel in Ireland than to Produce Biodiesel from Indigenous Irish Rape Seed?, *Applied Energy*, Vol. 86, No. 5, pp 595-604, 2009.

23. Clancy, M., et al., Bioenergy Supply Curves for Ireland 2010 – 2030, 2012, SEAI & AEA.

24. Murphy, J. D., et al., The Resource of Biomethane, produced via Biological, Thermal and Electrical Routes, as a Transport Biofuel, *Renewable Energy*, Vol. 55, No. 0, pp 474-479, 2013.

25. Howes, P., et al., UK and Global Bioenergy Resource, Report to DECC, AEA, 2011.

26. Ó Gallachóir, B. P., et al., Irish TIMES Energy Systems Model, in EPA Climate Change Research Programme 2007-2013, Report Series No. 242012, UCC, Johnstown Castle, Co.Wexford, Ireland, http://erc.epa.ie/safer/reports

27. Gargiulo, M. and Gallachóir, B. O., Long-term Energy Models: Principles, Characteristics, Focus, and Limitations, *Wiley Interdisciplinary Reviews: Energy and Environment*, Vol. 2, No. 2, pp 158-177, 2013.

28. Loulou, R., et al., Documentation for the TIMES Model, Energy Technology Systems Analysis Programme (ETSAP), 2005, http://www.etsap.org/documentation.asp

29. IEA-ETSAP, Joint Studies for New and Mitigated Energy Systems, Final Report of Annex XI (2008-2010), Vaillancourt, K., and Tosato, G., Editors, 2011.

30. IEA-ETSAP, Global Energy Systems and Common Analyses, Final Report of Annex X (2005-2008), ed. Goldstein, G., and Tosato, G., 2008.

31. Chiodi, A., et al., Modelling the Impacts of Challenging 2050 European Climate mitigation Targets on Ireland's Energy System, *Energy Policy*, Vol. 53, No. 0, pp 169-189, 2013.

32. Fitzgerald, J., et al., Medium-Term Review: 2013-2020., in No. 12, ESRI Forecasting Series, ESRI: Dublin, Ireland, 2013.

33. IEA, World Energy Outlook, Paris, France, IEA Publications, 2012.

34. SEAI, Bioenergy Roadmap, 2010.

35. Phillips, H., All Ireland Roundwood Production Forecast 2011-2028., COFORD, Department of Agriculture, Fisheries and Food, Dublin, Ireland, 2011.

36. McEniry, J., et al., The Effect of Feedstock Cost on Biofuel Cost as Exemplified by Biomethane Production from Grass Silage, *Biofuels, Bioproducts and Biorefining*, Vol. 5, No. 6, pp 670-682, 2011.

37. Kent, T., Kofman, P. D. and Coates, E., Harvesting Wood for Energy Cost-effective Woodfuel Supply Chains in Irish Forestry, 2011, http://www.coford.ie/media/coford/content/publications/projectreports/Harvesting_Wood_low_res_for_web.pdf

38. Clancy, D., et al., The Economic Viability of Biomass Crops versus Conventional Agricultural Systems and its Potential Impact on Farm Incomes in Ireland, in 107th EAAE Seminar, Modelling of Agricultural and Rural Development Policies, January 29th-February 1st, Sevilla, Spain, 2008.

39. RES2020, http://www.cres.gr/res2020/, [Accessed: May-2013]

40. Parsons Brinckerhoff, *Electricity Generation Cost Model - 2011 Update, Revision 1*, Prepared for Department of Energy and Climate Change: London, UK, 2011.

41. VGB Powertech, *Investment and Operation Cost Figures – Generation Portfolio, Survey 2011*, Essen, Germany, 2011.

42. Parsons Brinckerhoff, *Solar PV cost update*, Prepared for Department of Energy and Climate Change: London, UK, 2012.

43. EirGrid and SONI, All Island TSO facilitation of renewables Studies, Final Report, http://www.eirgrid.com/renewables/facilitationofrenewables/2010

44. EPA, Ireland's Greenhouse Gas Emissions Projections, 2012-2030, Environmental Protection Agency, 2013.

45. Glynn, J., et al., Energy Security Analysis: The Case of Constrained Oil Supply for Ireland, *Energy Policy*, Vol. 66, No. 0, pp 312-325, 2014.

46. EC, COUNCIL REGULATION (EC) No 73/2009 of 19 January 2009 establishing Common Rules for Direct Support Schemes for Farmers under the Common Agricultural Policy and establishing Certain Support Schemes for Farmers, amending Regulations (EC) No 1290/2005, (EC) No 247/2006, (EC) No 378/2007 and repealing Regulation (EC) No 1782/2003, *Official Journal of the European Union*, 2009.

47. EC, COMMISSION REGULATION (EC) No 1122/2009 of 30 November 2009 laying down Detailed Rules for the implementation of Council Regulation (EC) No 73/2009 as Regards Cross-compliance, Modulation and the Integrated Administration and Control System, under the direct Support Schemes for Farmers provided for that Regulation, as well as for the implementation of Council Regulation (EC) No 1234/2007 as Regards Cross-compliance under the Support Scheme provided for the Wine Sector, *Official Journal of the European Union*, 2009.

48. SEAI, Bioenergy Atlas, http://maps.seai.ie/bioenergy/

49. Smyth, B. M., Murphy, J. D. and O'Brien, C. M., What is the Energy Balance of Grass Biomethane in Ireland and other Temperate Northern European Climates?, *Renewable and Sustainable Energy Reviews*, Vol. 13, No. 9, pp 2349-2360, 2009.

50. McEniry, J., et al., How much Grassland Biomass is Available in Ireland in excess of Livestock Requirements?, *Irish Journal of Agricultural and Food Research*, Vol. 52, No.1, pp 67-80, 2013.

7

The Potential of a Hybrid Power Plant for the Dubrovnik - Neretva County (Southern Croatia)

Huili Zhang[*1], *Jan Baeyens*[2], *Jan Degreve*[3]

[1]Department of Chemical Engineering, Chemical and Biochemical Process Technology and Control Section, KU Leuven, Heverlee, Belgium
e-mail: Zhanghl.lily@gmail.com
[2]College of Life Science and Technology, Beijing University of Chemical Technology, Beijing, China
[3]KU Leuven, Heverlee, Belgium

ABSTRACT

The Croatian electricity demand exceeds domestic production and about 30% of additional power is covered from imports. The Croatian government is planning to add domestic production capacity, using natural gas and coal as the main fuel. Due to the attractive solar irradiation in Croatia, these new projects offer the possibility to build these thermal power plants as hybrid plants by adding solar energy within the production. The present paper will investigate the potential of solar energy in the Southern Croatian region and its possible contribution in a hybrid set-up. The paper assesses the required input data and presents a hybrid Concentrated Solar Power (CSP) design. The results demonstrate the potential of adding a CSP in a hybrid power plant in Southern Croatia.

KEYWORDS

Croatia, Concentrated solar power, Design, Hourly direct irradiation, Plant simulation.

INTRODUCTION

Croatia has a total installed electricity generation capacity of 3,763 MW, including 2,097 MW of hydropower and 1,666 MW from thermal power plants (including a nuclear power plant, jointly owned with Slovenia) [1]. The domestic demand exceeds domestic production and about 30% of additional power is covered from imports. The Croatian government intends to restructure, liberalize and privatize the energy sector to emphasize compatibility with the European Union, whilst also planning to increase the domestic production capacity, especially with natural gas and coal as the fuel [2]. For Southern Croatia, the installation of an 800 MW coal-fired power plant at Ploče is proposed, albeit severely opposed by eco-activists [3, 4]. The main concerns relate to the potential of the Neretva river valley for ecological food production, to the possible pollution of the river, to the emission of hazardous pollutants (mainly fine dust) and to the negative impact of such a power plant upon the tourism in the region. The thermal power plant will moreover necessitate the transport of hundreds of tons of coal per day. Although modern coal-fired thermal power plants implement highly efficient techniques for dust abatement and for the reduction of common pollutants (e.g. SO_2), the CO_2 emissions of a coal-fired power plant remain a major issue. Unless an expensive CO_2 capture is foreseen and depending on the quality of the coal, CO_2 emissions will vary between 0.5 and 1 ton CO_2/MWh.

---

* Corresponding author

As elsewhere, public concern, mostly about adding coal-fired power plants, has generated uncertainty in the electricity market, whilst creating a very favorable context towards new Renewable Energy projects, with a focus on using solar energy [5-7].

In March 2007, the European Union targeted 20% renewable energy for 2020, with special emphasis on small scale units.

Due to the attractive solar irradiation in Southern Croatia, new power plant projects offer the possibility to build them as hybrid plants by adding solar energy within the production. Such a hybrid solution, where solar energy will partly replace the coal resource, will not only reduce the pollutant emissions (including CO_2), but will moreover create an additional tourist attraction pole, as is the case with the Gemasolar solar power plant in Southern Spain [8]. An additional advantage results from the fact that the power block (boiler, turbines) will be available within the thermal power plant, thus only the investment in heliostats, solar receiver and heat carrier circuits need to be considered.

The present paper investigates the potential of solar energy in the Southern Croatian Dubrovnik-Neretva county and its possible contribution in a hybrid set-up. Such a hybrid solution of a traditional fossil fuel fired thermal power plant with a supply of solar energy during the daytime and during the night time if thermal energy storage is foreseen, will certainly create a more positive assessment by the public in comparison with a sole fossil fuel power plant.

To enhance the rate of this solar energy development, it is necessary to update the insights, the tools and the technical/economic analysis. Within the solar energy technologies, photovoltaics (PV) to a large extent and Concentrated Solar Power technology to a lesser extent, have been widely investigated and applied. PV draws a significant focus with a guaranteed future in view of the ongoing technical improvements and cost reduction [7, 9, 10]. Concentrated Solar Power Plants are gaining increasing interest, mostly by using the Parabolic Trough Collector (PTC) system, but with Solar Tower Collectors (STC) progressively occupy a significant market position. The large-scale STC technology was successfully demonstrated by Torresol in the Spanish Gemasolar project on a 19.9 MW_{el} scale [8]. The varying solar radiation flux throughout the day and throughout the year remains a main problem for all solar energy technologies. A more efficient operation requires the incorporation of two technologies, i.e. Thermal Energy Storage (TES) to cover non-sunny periods and Backup Systems (BS) or Hybrid Operation when solar energy or TES supply are not available. The combination of both systems facilitates a successful continuous and year round operation. To determine the optimum design and operation of the CSP throughout the year, whilst additionally defining the capacity of TES and required BS, an accurate estimation of the daily solar irradiation is needed, accounting for hourly irradiation fluxes and direct irradiation, since a CSP plant will only accept Direct Normal Irradiance (DNI) in order to operate.

Most of the countries, except those above latitude 45 °N or below latitude 45 °S, are subject to an annual average irradiation flux in excess of 1.6 MWh/m² [6]. The procedure, outlined in the present paper, combines previous theoretical and experimental findings into a general method of calculating the hourly beam (direct) irradiation flux and applying it to the design of CSP applications.

The present paper will hence:

- Briefly review the CSP technologies and their advantages;
- Estimate the hourly beam irradiation flux from available monthly mean global irradiation data for 2 selected European locations, being Sevilla as reference for the Gemasolar project [8] and Dubrovnik, as reference for the Dubrovnik-Neretva County;
- Select an appropriate plant configuration and perform a preliminary design;

- Estimate the Levelized Electricity Cost for the hybrid potential in Southern Croatia.

CSP TECHNOLOGIES

Concentrated Solar Power (CSP) generates electricity by using heat provided by concentrated solar irradiation. Zhang *et al.* [6] reviewed available CSP technologies, including Parabolic Trough Collector (PTC), Solar Power Tower (SPT), Linear Fresnel Reflector (LFR), Parabolic Dish Systems (PDS) and Concentrated Solar Thermo-electrics (CSTe). Sunlight is reflected to a receiver (either a tower in SPT or focus heat capture pipes in PTC), where heat is collected by a thermal energy carrier (primary circuit) and mostly used via a secondary circuit to power a turbine. PTC and SPT have been built and developed around the world since 1981, with capacities ranging from 0.5 MW to more than 300 MW [6]. Additional large-scale CSPs (2014-2017) will have capacities of 100's MW_{el}.

Short-term, direct Thermal Energy Storage stores excess heat collected in the solar field by the Heat Transfer Fluid (HTF), thus avoiding losing the daytime surplus energy while extending the production after sunset. Long term thermal energy storage is less obvious, since it involves storage covering months of low solar irradiation. Currently, only sensible heat is stored. The significant improvement by using latent heat storage (phase change materials) or even chemical heat storage (reversible endothermic/exothermic synthesis) is under development [11-13]. In current TES, liquids (oils, molten salts) are used for sensible heat capture and storage. Indirect storage uses solid heat absorbers and a secondary HTF circuit. SPT plants commonly use a BS to regulate production and guarantee a nearly constant capacity. The reference Gemasolar CSP uses molten salts (<565 °C), a steam boiler and advanced Rankine cycle turbine. After discounting the parasitic energy use (mostly molten salt circulation, storage and boiler), the resulting conversion efficiency is about 16%. With novel HTF materials, such as powders, the temperature range will be extended to well above 750 °C, with reduced parasitic energy use and an expected overall conversion around 20% [14-17]. The measures to enhance the efficiency of CSP plants are in full development. The Ploče initiative could certainly benefit from the expected conversion increase of 20 to 25%.

COMPUTING GLOBAL AND DIRECT SOLAR HOURLY IRRADIATION

Calculation method

The equations and sequence of the calculation method were presented by Zhang *et al.* [6] and applied to the design of a CSP plant in Chile [18] and to the assessment of current photovoltaic initiatives [7]. The essential steps involve known data, general assumptions and a calculation sequence. The basic data include:

- The angle and distance of the sun vs. position on earth as function of latitude and time;
- Satellite data of monthly average solar irradiation H [19];
- Data on temperatures, rainfall, wind speed;
- Day of the year, sunrise/sunset time. The calculation sequence involves calculating:
  - The solar extra-terrestrial irradiation, H_0;
  - The average clearness index $K_{T,av}$ (= H/H_0, or via Hargreaves [20]), daily K_T, and daily total irradiation = $K_T H_0$;
  - The sequence of days (not needed with Hargreaves K_T);
  - The daily diffuse irradiation H_d as fraction of H;

○ Predicting the hourly values of total I/H, diffuse I_d/H and direct (beam) irradiation $I_b = I - I_d$.

Model parameters and selected locations

There are two reliable sources that provide information on essential meteorological parameters: monthly mean temperature and solar radiation. These sources are the NASA website [19] with respect to solar radiation data and TUTIEMPO [21] to provide daily mean, maximum and minimum temperature data for any given location. The NASA data are available on a mean-monthly basis, whereas TUTIEMPO are downloadable on a day-by-day basis. It is important to remember that NASA data are based on satellite observations that represent inferred values of irradiation; in contrast, TUTIEMPO provides ground-measured data for temperature. Hence, if reliable regressions are available between irradiation and mean temperature, then the latter data may be used to obtain more realistic estimates of irradiation [6].

To illustrate the use of the methodology, two European locations were selected, i.e. Sevilla as a reference, since it is close to the Gemasolar STP and Dubrovnik as focal point of the Dubrovnik-Neretva County. The essential data of the locations are given in Table 1.

Table 1. Selected locations with basic data [19, 21]

Location	Latitude [rad]	Longitude [rad]	$\bar{H}$ (January) [kWh/m² day]	$\bar{H}$ (July) [kWh/m² day]	January		July	
					T_{max} [°C]	T_{min} [°C]	T_{max} [°C]	T_{min} [°C]
Sevilla, Spain	37.41	-5.98	2.56	7.80	15.9	7.6	35.7	19.9
Dubrovnik, Croatia	42.65	18.09	1.61	5.96	10.1	2.9	31.3	22.2

RESULTS AND DISCUSSION

Direct irradiation

The monthly extra-terrestrial irradiation, H_0 is a function of the latitude only, and for Dubrovnik ranging from ~11.5 kWh/m² day in June to ~3.8 kWh/m² day in December.

To proceed with the calculation of the monthly average clearness index, $\overline{K_T}$, H_0 is used together with NASA data [19]. Results are illustrated as example in Figure 1 for the Dubrovnik location.

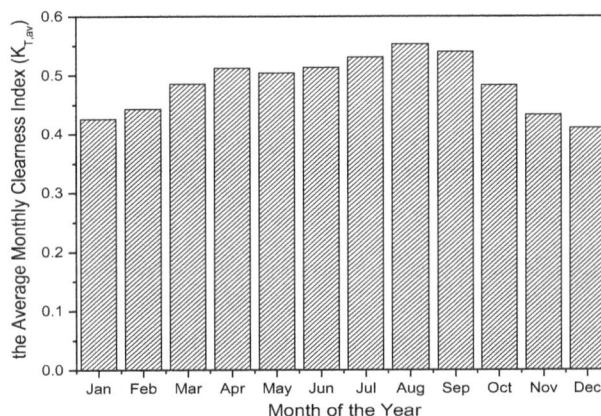

Figure 1. Calculated monthly average clearness index for Dubrovnik

The daily total irradiation is thereafter obtained by applying the daily clearness index K_T. Figure 2 shows the model-predicted total daily irradiation, ordered in ascending daily pattern for Dubrovnik for a summer month (July) and a winter month (January). The monthly average H in January and July, are 1.67 kWh/m² day and 5.92 kWh/m² day, respectively. Similar trends are obtained for Sevilla.

Applying the sequence model for the daily clearness indexes, as function of K_T [6], transforms the ascending nature of the consecutive days into a wave-function, although monthly average values of H remain unchanged.

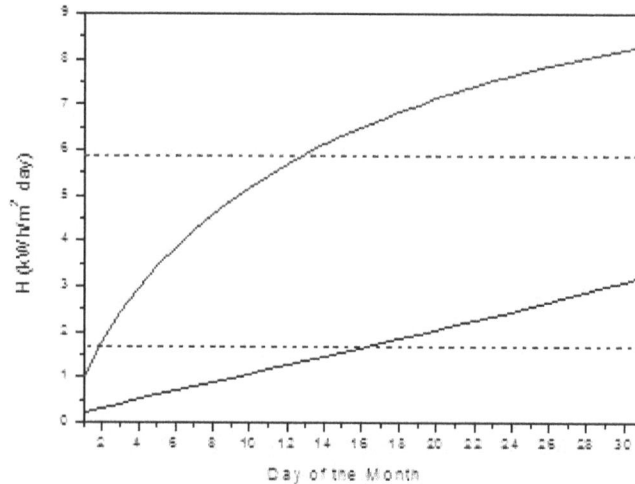

Figure 2. The total daily irradiation in Dubrovnik: - - - average; — ascending pattern (- - - — in January and - - -, — in July)

The most important result towards CSP design requires the direct (beam) irradiation, obtained by subtracting the diffuse irradiation, H_d, from the total irradiation H. The ratio of the diffuse to total irradiation in Dubrovnik is ~0.5 in August and ~0.66 in December.

The resulting beam radiation H_b, as detailed daily/monthly average values for Dubrovnik and reference Sevilla, is shown in Figure 3.

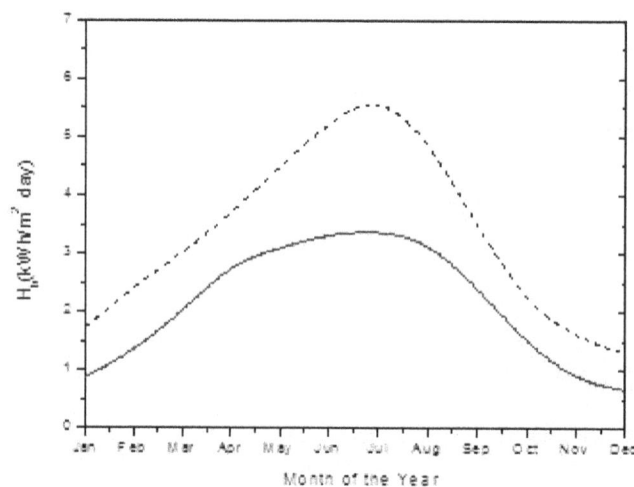

Figures 3. Average monthly direct (beam) irradiation: - - - in Sevilla; — in Dubrovnik

Due to its lower latitude, Sevilla presents much higher values of H_b than Dubrovnik.

Finally, a complete hourly profile can be predicted by the model, as illustrated in Figure 4, where the radiation flux can be seen to increase from sunrise to noon, and thereafter decreasing again till sunset.

Figure 3 also implies that the selection of the CSP nominal capacity will be a compromise between the seasons, accounting for the capability of thermal storage, and the use of a backup system.

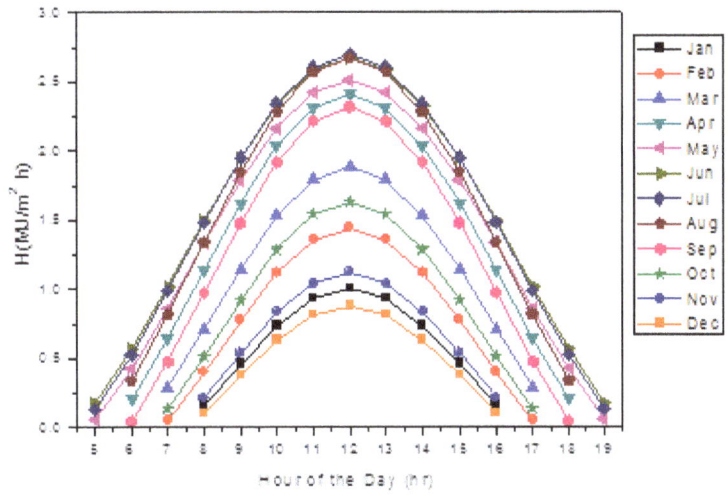

Figure 4. Hourly values on the 15th of the respective months in Dubrovnik

Methodology to apply the predictions in CSP design

Having established the annual, monthly and daily levels of direct (beam) solar irradiation, its impact on the power yield of the CSP plant can be assessed.

To do so, it should be remembered that the overall CSP-layout has its overall efficiency. The projected overall efficiency of CSP plants currently exceeds 16% and is expected to reach ≥20% by 2020 [6, 18, 22]. The current efficiencies of the essential components have been reported by Sargent and Lundy Consulting Group [22].

Considering the application of the Solar Tower Collector, with molten salt HTF/TES and with natural gas or coal-fired BS, and assuming the use of a heliostat field (HFC) of 318,000 m² (as applied by Gemasolar), Figure 5 provides some indications of the simulated results.

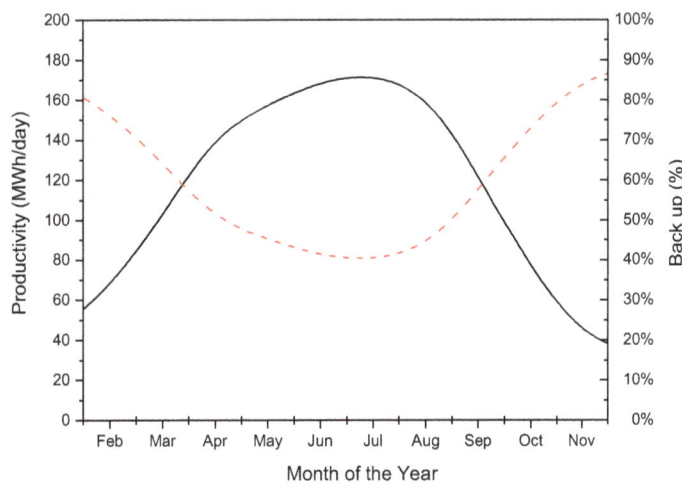

Figure 5. Electricity Generation (at an overall 16% conversion) throughout the year in the possible Dubrovnik-Neretva SPT, in hybrid operating mode: — solar production; - - - back up requirements

It is understood that the annual shut-down/maintenance period will be between December 15th and January 15th. The results of the simulation for the Dubrovnik-Neretva initiative reveal that the solar generation will account for ~36 GWh/year, in the case of a conservative 16% overall efficiency. At 20% efficiency, as can be expected for the Dubrovnik-Neretva plant using newest technologies, the production will increase to 45 GWh/year. The back-up requirements in the Dubrovnik-Neretva case are intrinsically available through the coal-fired thermal power plant.

In locations with higher solar irradiation, e.g. Gemasolar [8] or Calama [18], the solar energy contribution will be proportionally enhanced.

To significantly increase the solar contribution for the Dubrovnik-Neretva power plant, the size of the HFC will need to be increased.

Levelized Energy Cost

A further demonstration of the SPT potential for Dubrovnik-Neretva is provided by a tentative Levelized Energy Cost (LEC) calculation, commonly used to compare competing energy sources and expressed in USD/kWh. The total investment includes the costs of the solar field and of the TES. Since the proposed plant will be of hybrid nature, the costs related to the coal-fired BS and associated cooling process, were not included in the calculation.

Economic factors from literature were used, as summarized in Càceres *et al.* [18]. A 10% discount rate and 25 years life time of the plant were used.

At a pessimistic overall efficiency of 16%, LEC-values for the sole solar energy contribution vary between 0.10 and 0.12 USD/kWh. An enhanced overall efficiency (16 to 20%) will reduce the LEC proportionally. These LEC results are consistent with literature references for solar energy prices. The technical and economic potential of the Dubrovnik-Neretva hybrid SPT project is certainly significant, and reduces the environmental burden of a sole coal-fired power plant. Costs of solar field and power block are moreover decreasing, which will positively affect the LEC-value. The introduction of powder circulation heat carriers and Phase Change Materials is also expected to significantly reduce the LEC [12, 13, 16].

CONCLUSIONS

The paper develops the underlying equations to calculate the daily total irradiation, the direct (beam) and diffuse irradiation. Having established the annual, monthly and daily levels of direct (beam) solar irradiation, its impact on the power yield of the CSP can be assessed. The projected overall efficiency of CSP plants was included in a CSP performance simulation. Initial simulation results are illustrated for a 12 MW_{el} Solar Power Tower project, with molten salts as HTF and operating in a hybrid way with the projected coal-fired thermal power plant at Ploce, in the Dubrovnik-Neretva County. In the assessed example, solar generation will account for ~45 GWh/year, with possible higher yields due to improvements in HFC, solar tower design and reduced parasitic electricity consumption. The results demonstrate the potential of adding a CSP in a hybrid operation of the coal-fired power plant of Ploče.

NOMENCLATURE

H_0	Extra-terrestrial radiation	[kWh/m^2 day]
H	Daily total radiation obtained from the registered measurements	[kWh/m^2 day]
$\bar{H}$	Monthly average of H	[kWh/m^2 day]
I, I_d, I_b	Hourly solar radiation, with diffuse and beam components, respectively	[MJ/m^2 h]

| $K_{T,av}$ | Monthly average clearness index | [-] |
| K_T | Daily clearness index | [-] |

Abbreviations

BS	Backup System
CSP	Concentrated Solar Power Plant
CSTe	Concentrated Solar Thermo-electrics
DNI	Direct Normal Irradiance
HFC	Heliostat Field Collector
HTF	Heat Transfer Fluid
LEC	Levelized Electricity Cost
LFR	Linear Fresnel Reflector
PDS	Parabolic Dish System
PTC, STC	Parabolic Trough Collector; Solar Tower Collector, respectively
PV	Photovoltaics
SPT	Solar Power Tower
TES	Thermal Energy Storage

REFERENCES

1. HEP, Croatia, http://www.hep.hr/ops/en/hees/data:aspx
2. National Energy Grid Croatia, www.geni.org/globalenergy/librang/national_energy_grid/Croatia/EnergyOverviewofCroatia
3. Croatian Times, Eco Activists oppose Plans to build a Power Plant near Ploce, www.croatiantimes.com, 05 June 2014.
4. Neretva Valley, Metkovic Mayor against Thermal Power Plant in Neretva Valley, www.dalje.com, [Accessed: 01-March-2014]
5. OECD/IEA, Technology Roadmap, Concentrating Solar Power, 2010.
6. Zhang, H. L., Baeyens, J., Degrève, J., Cacères, G., Concentrated Solar Power Plants: Review and Design Methodology, *Renewable and Sustainable Energy Reviews*, Vol. 22, pp 466-481, 2013.
7. Zhang, H. L., Van Gerven, T., Baeyens. J. and Degrève, J., Photovoltaics: Reviewing the European Feed-in-Tariffs and Changing PV Efficiencies and Costs, *The Scientific World Journal*, 404913, pp 1-10, 2014.
8. http://www.torresolenergy.com/TORRESOL/gemasolar-plant/en
9. EPIA, Connecting the Sun, Solar Photovoltaics on the Road to Large-Scale Grid Integration, September 2012.
10. Tyagi, V. V., Rahim, N. A. A., Rahim, N. A., Selvaraj, J. A. L., Progress in Solar PV Technology: Research and Achievement, *Renewable and Sustainable Energy Reviews*, Vol. 20, pp 443-461, 2013.
11. Fernandes, D., Pitié, F., Cáceres, G., Baeyens, J., Thermal Energy Storage "How previous findings determine current research priorities" *Energy*, Vol. 39, No. 1, pp 246-257, 2012.
12. Pitié, F., Zhao, C. Y., Baeyens, J., Degrève, J., Zhang, H. L., Circulating Fluidized Bed Heat Recovery/storage and its Potential to use Coated Phase-change-material (PCM) Particles, *Applied Energy*, Vol. 109, pp 505-513, 2013.
13. Zhang, H. L., Baeyens, J., Degrève, J., Cáceres, G., Segal, R., Pitié, F., Latent Heat Storage with Tubular-encapsulated Phase Change Materials (PCMs), *Energy*, (in press), 2014.

14. Zhang, H. L., Baeyens, J., Degrève, J., Brems, A., Dewil, R., The Convection Heat Transfer Coefficient in a Circulating Fluidized Bed (CFB), *Advanced Powder Technology*, Vol. 25, No. 2, pp 710-715, 2014.

15. Zhang, H. L., Flamant, G. , Gauthier, D., Ansart, R., Hemati, M., Baeyens, J., Boissière, B., The use of Dense Particle Suspensions as Heat Transfer Carrier in Solar Thermal Plants, International Conference on Solar Heating and Cooling, pp 25-27, June 2014, Gleisdorf, Austria.

16. Flamant, G., Gauthier, D., Benoit, H., Sans, J. L., Garcia, R., Boissière, B., Ansart, R., Hemati, M., Dense Suspension of Solid Particles as a new Heat Transfer Fluid for Concentrated Solar Thermal Plants: On-sun Proof of Concept, *Chemical Engineering Science*, Vol. 102, pp 567-576, 2013.

17. Barlev, D., Vidu, R., Stroeve, P., Innovation in Concentrated Solar Power, *Solar Energy Materials & Solar Cells*, Vol. 95, No. 10, pp 2703-2725, 2011.

18. Cáceres, G., Anrique, N., Girard, A., Degrève, J., Baeyens, J., Zhang, H. L., Performance of molten Salt Solar Power Towers in Chile, *Journal of Renewable and Sustainable Energy*, Vol. 5, 053142, 2013.

19. NASA, http://eosweb.larc.nasa.gov/cgi-bin/sse/retscreen.cgi?email=rets@nrcan.gc.ca

20. Hargreaves, G., Simplified Coefficients for estimating Monthly Solar Radiation in North America and Europe, Dept. paper, Department of Biology and Irrigation Engineering, Utah State University, 1994.

21. Tutiempo, http://www.tutiempo.net/en/

22. Sargent and Lundy Consulting Group, Assessment of Parabolic Trough and Power Tower Solar Technology Cost and Performance Forecasts, National Renewable Energy Laboratory, 2003.

8

Future Generation Adequacy of the Hungarian Power System with Increasing Share of Renewable Energy Sources

Agnes Gerse

Department of Energy Engineering, Budapest University of Technology and Economics,
MAVIR ZRt., Budapest, Hungary
e-mail: gerse@mavir.hu

ABSTRACT

The power generation sector is expected to undergo substantial changes in Hungary in the near future due to the decommissioning of several large units reaching the end of their lifetimes in parallel to the projected increase of renewable electricity generating capacity. In addition to the traditionally widely used deterministic adequacy assessment methods, a probabilistic approach has a great importance in case of technologies with different capacity credits. An analytical country-specific adequacy assessment model enabling the probabilistic modelling of wind power plants was developed and applied to generating capacity forecasts for Hungary. Model parameters were estimated using multi-annual production, plant availability and hourly system demand data. Adequacy indicators obtained from the model clearly show increasing reliance on imported electricity in the absence of investments in new generating capacity.

KEYWORDS

Generation adequacy assessment, Security of supply, Reliability, Demand, Power generation, Renewable Energy Sources (RES), Prospective system analysis.

INTRODUCTION

Generation adequacy is a key element of electric system reliability, ensuring that there is sufficient installed capacity in the power system to meet the electric load [1], including reserve capacity to perform corrective and preventive maintenance, as well. At present, a wide range of generation adequacy assessment methodologies and adequacy metrics are used in Europe [2], that were originally developed for thermal and hydro dominated systems. However, there is a growing expectation for the improvement of methodology due to the increasing share of RES production of highly stochastic nature. In parallel with RES integration, the absence of investments and premature decommissioning of existing units can lead to a shortfall in conventional generating capacity, as well. Therefore electricity transmission system operators have a very important role in contributing to the monitoring of security of supply in order to provide relevant information for decision-makers.

Despite of the considerable work done on the development of generating capacity reliability assessment, the dominant practice has been to use deterministic approach with deterministic adequacy criteria, such as a reserve capacity equal to a percentage of the expected load. Recently, probabilistic methods have gained more and more importance because of their ability to respond to stochastic factors influencing system reliability. Probabilistic criteria and indices include LOLP, LOLE, LOEE, EENS and EUE. Both analytical approach (mathematical models using mathematical solutions) and simulation techniques are applied for the evaluation of reliability indices [3].

In the last two decades, new methods were introduced to enable the capacity adequacy evaluation of power systems with a significant share of wind power plants. Because of the limited availability of normalised wind power production time series, site-specific wind data simulation models were developed and validated consisting of wind speed models and wind turbine generation models based on power curves [4]. As a simplification, a multi-state approach was proposed for wind farms providing sufficient accuracy by considering a large number of discrete output states [5, 6]. Based on the multi-state approach, D'Annunzio and Santoso [7, 8] used multiple years of wind power output data to construct the capacity outage individual probability table for wind power plants. More recently, research has focused on the capacity credit calculation methods of wind power plants [1, 7, 9].

In this paper, the multi-state representation is integrated into a generation adequacy assessment model elaborated Hungarian power system. Despite of the limitations of a one-country model, which should be kept in mind when considering the increasing role of cross-border exchanges, it can provide a substantial contribution for prospective analysis at national level for quantifying the reliance on imported electricity.

SYSTEM CHARACTERISTICS, ADEQUACY CRITERIA

Since the paper presents a country-specific adequacy assessment model, the main system characteristics and adequacy criteria are briefly summarized prior to the description of methodology.

The present electricity generation mix of Hungary is dominated by nuclear and thermal power plants. The power generation portfolio consists mostly of large power plants (with a generating capacity above 50 MW at each plant), the majority of them were commissioned before 1990. From the beginning of the 2000s, a significant share of power capacity addition was small-scale CHP to supply local heat demand (district heating, industry, public institutions, etc.).

The main country statistics of 2012 are summarised in Table 1. Almost half of the domestic electricity production came from nuclear power (Paks Nuclear Power Plant). Other power plants using fossil fuels (mostly gas and lignite) supplied more than 40%.

Table 1. Capacity and production data of Hungary (2012) [11]

	Net generating capacity (as of 31 December 2012)		Net annual production	
	[MW]	[%]	[TWh]	[%]
Nuclear	1,892	20.7	14.8	46.4
Lignite	740	8.1	4.8	15.1
Hard coal	294	3.2	0.8	2.5
Gas	5,229	57.2	9.0	28.2
Oil	410	4.5	0.1	0.3
Hydro	56	0.6	0.2	0.6
Wind	329	3.6	0.7	2.2
Solar	0.4	0.0	0.0	0.0
Biomass	189	2.1	1.5	4.7
Total	9,139	100	31.9	100

Because of geographical constraints, hydro power plants (all of them run-of-river, without storage capability) have a marginal role in domestic power generation. However, non-hydro RES production has a growing importance due to the commitment to sustainability. Considering the moderate wind [10] and solar generating potential of Hungary, RES production is based mostly on biomass, including biomass co-firing in coal fired power plants. The net wind generating capacity reached 329 MW by the end of 2012. Solar PV installations have a negligible total generating capacity at the moment. (The vast majority of solar PV generation is connected to the grid as embedded generation.)

As a well-interconnected country, cross-border resources can provide a significant contribution to the electricity supply of Hungary subject to availability, heavily influenced by market conditions. Hungary has been a net importer of electricity for several years. Recently, a significant increase in electricity import has been observed. In addition to the 31.9 TWh of domestic generation, 8 TWh of electricity was imported in 2012 in order to cover the total consumption of 39.9 TWh. Net peak load of the system reached 6,016 MW in 2012, while the minimum of load was 2,586 MW.

After liberalisation of the electricity market, there is no more centralized capacity expansion planning in place in Hungary. However, mid- and long-term generation adequacy assessment reports are published by the transmission system operator on a regular basis. This adequacy assessment is based on a deterministic methodology, calculating system capacity balances for the peak load and assessing the level of expected remaining capacity. According to the former UCTE methodology, remaining capacity is the excess capacity above the forecasted peak load, taking into account expected unavailable capacity, outage, maintenance and system operation reserve.

In addition to the deterministic adequacy assessment approach, also a probabilistic generation adequacy standard exists: a LOLP based security of supply criterion is stated by the Grid Code. According to that criterion, LOLP cannot exceed 1% at any time. No methodology is specified by the Grid Code (e.g. clarification on the assessment of cross-border resources, capacity credit of wind power plants). Usually the results of a LOLP calculation for the projected annual peak loads are presented in the generation adequacy reports.

As shown by Figure 1 prepared using past retrospective capacity balances of 2011, the most constrained months are the summer months. Therefore mid- and long-term probabilistic adequacy analysis cannot be limited to the expected winter peak loads.

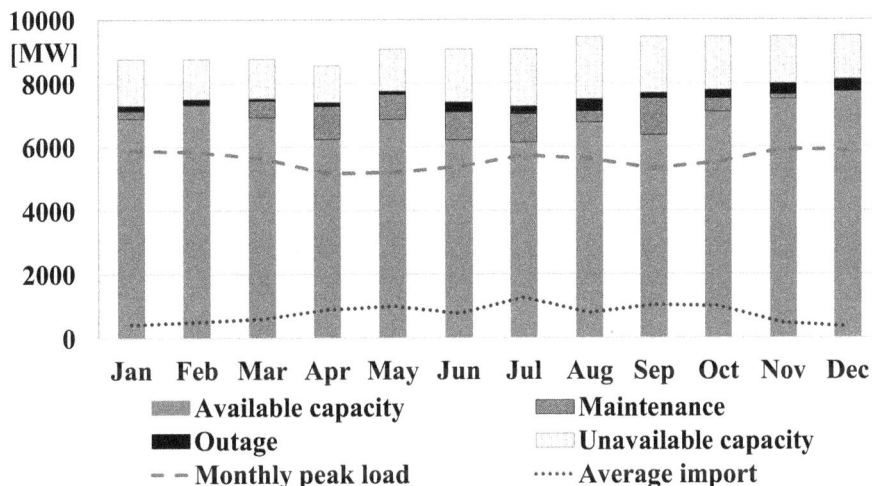

Figure 1. Monthly peak load vs. available net generating capacity in 2011 [12]

THE ADEQUACY ASSESSMENT MODEL

An adequacy assessment model was developed in order to enable an extended probabilistic adequacy assessment taking into account RES with weather-dependent generation patterns, constraints resulting from maintenance as well as co-generation (CHP). The model can be applied for the calculation of different adequacy indicators: LOLP, LOLE (h) and LOLE (d).

Supply side

In the adequacy assessment model, net generating capacity is divided into the following categories: large conventional power plants, hydro power plants, wind power plants and other generating capacity (including small-scale CHP and solar PV generating capacity). Adequacy indicators are computed using an analytical method. Depending on the power plant categories, two-state models, multi-state models and aggregated firm capacity are used for the calculation of COPT.

Large conventional power plants. Large power plants with a nominal generating capacity of above 50 MW are modelled per unit using a two-state model. In the case of existing units, FOR and maintenance requirements were estimated on the basis of past plant availability data (period of 1990-2008) and retrospective system balances [12]. For prospective analysis, assumptions on the plant availability data of projected new units have to be made, as well.

Depending on the technology, the temperature sensitivity of the output power and the effects of the heating period were also considered. (In Hungary, heating period is usually between 15 October and 15 April.) De-rated capacities were assumed for a number of CHP CCGT units in the summer months, since these can be operated with auxiliary cooling only when heat demand is low. Beyond that, the available generating capacity of non-CHP units is also influenced by ambient air temperature due to changes in gas turbine inlet air temperatures, cooling water temperatures, etc. In order to reflect these seasonal effects, nominal net generating capacities were multiplied by factors expressing the ratio of available and nominal capacity. These factors were estimated on a monthly basis based on past plant operation data.

Hydro power plants. For the assessment of the relatively small total generating capacity of run-of river hydro power plants, average monthly capacity factors can be used. These capacity factors can be calculated from past monthly production data [11]. Table 2 contains average monthly values for period 2002-2009, for wet, average and dry years.

Table 2. Average monthly capacity factors of Hungarian hydro power plants

	Jan	Feb	Mar	Apr	May	Jun	Jul	Aug	Sep	Oct	Nov	Dec
Dry year	0.23	0.11	0.13	0.10	0.28	0.13	0.30	0.24	0.20	0.40	0.38	0.34
Average year	0.36	0.37	0.22	0.42	0.47	0.41	0.46	0.42	0.46	0.48	0.43	0.42
Wet year	0.43	0.47	0.34	0.6	0.56	0.58	0.53	0.58	0.53	0.58	0.49	0.46

Wind power plants. In order to assess the weather-dependent production patterns of wind power plants, a multi-state model was elaborated based on the approximate method proposed by D'Annunzio and Santoso [7, 8]. This method assumes that wind production can be properly modelled by a large number of discrete output states (partial capacity outage states) for adequacy assessment purposes. Instead of using synthetic data, the method is based on power output data.

The 15-minute resolution time series of total wind production in Hungary were collected as input from the years 2011 and 2012. Because of the intra-annual increase of the total installed wind generating capacity, time series were normalised and load factors were calculated. As an approximation, the load factor time series can be used for projected higher net generating capacities, but limitations (e.g. smoothing effect, influence of geographical distribution) should be kept in mind. In addition, aggregated on-line measurement data are also subject to errors, effects of maintenance, outages, trial runs under commissioning period, etc.

The multi-state representation of wind (Table 3) uses a load-factor resolution of 0.02, considering 51 capacity outage states. (This resolution is equivalent to 6.48 MW in case of the net installed capacity at the end of 2012.)

Table 3. Multi-state representation input data for wind power plants

Capacity outage state load factor (-)	Individual probability (-)
0.00	0
0.02	0.000228
0.04	0.000855
0.06	0.002280
0.08	0.003220
0.10	0.003434
...	...
0.90	0.040099
0.92	0.045172
0.94	0.052596
0.96	0.061431
0.98	0.084374
1.00	0.155324

Solar power plants. Due to the negligible present total net generating capacity, there are no aggregated online measurement data available for analysis. As a simplified approach, hourly production time series were estimated and taken into account on the basis of 2011 irradiation data. This approach can be justified by the moderate development targets (see Table 4).

Capacity Outage Probability Table (COPT)

COPT values express the cumulative probability $P(X \geq x)$ of having a system capacity outage greater than or equal to x. Values of COPT can be computed using a recursive algorithm [8]. When adding the new two-state unit $i+1$ with FOR of FOR_{i+1} and net generating capacity of C_{i+1} to the system containing already i units, the following formula can be used:

$$P(X_{i+1} \geq x) = (1 - FOR) \times P(X_i \geq x) + FOR \times P(X_i \geq x - C_{i+1}) \qquad (1)$$

In the case of multi-state modelling applied for wind power plants, considering m partial capacity outage states with individual probabilities of p_j and capacity outage states C_j, the capacity addition formula is as follows:

$$P(X_{i+1} \geq x) = \sum_{j=1}^{m} p_j \times P(X_i \geq x - C_j) \tag{2}$$

Since capacity outage states are given by load factors in Table 3, they need to be up-scaled according to the total net wind generating capacity to be considered.

In the model, capacity outage probability tables were calculated for large power plants, whose availability is described by a two-state model, and for wind power plants (considered as an aggregated multi-state unit) on a monthly basis. Unavailable capacity of hydro power plants, solar power plants and other small-scale units were taken into account as a fixed input.

Demand side

For the calculations presented here, past hourly system load time series were used, scaled up according to the forecasted annual demand. As well known, system load patterns exhibit annual, daily and weekly seasonality, and in addition to that, both long (e.g. economic framework) and short term effects (e.g. weathering conditions, calendar effects) have a strong influence on system load characteristics. For temperature sensitivity analysis purposes, also a statistical, multiple regression based load model is available.

The applicability of statistical models adjusted to present system load patterns will be very limited in case of a very long time frame; therefore new approach has to be elaborated in order to take into account the effects of demand side management, heat pumps, electric vehicles, etc., properly.

PROSPECTIVE ANALYSIS

In a liberalized market environment, the development of the power plant portfolio is an outcome of individual decisions taken by investors and power plant operators subject to market and regulatory framework. The expected technical lifetime of existing power generating assets can be indicative only for the expected time of decommissioning. Unfavourable market conditions and significant surplus capacity can lead to premature decommissioning or mothballing of units, as it can be observed in several European countries nowadays. In order to take into account the uncertainties, usually several scenarios are examined.

For prospective analysis, we considered the scenarios submitted by the Hungarian electricity transmission system operator for the ENTSO-E Scenario Outlook and Adequacy Forecast 2013-2030 (SO&AF) report (Figure 2) [13]. The ENTSO-E SO&AF report contains several scenarios for each member country. In Scenario A (conservative scenario), only firm new investments can be considered, while Scenario B is a best estimate scenario assuming that all necessary regulatory and economic incentives will be in place. It should be noted that considering the lead time required for thermal power plants, the year 2015 is very near. Due to delayed final investment decisions and/or cancellation of licensed projects, there are no large power plant units under construction in Hungary at the moment. This results in an identical net generating capacity forecast for Scenarios A and B in 2015. However, the 2020 forecasts reveal a large gap between the two capacity outlooks reflecting a conservative (firm capacity additions only) and a best estimate (necessary investments are in place) approach.

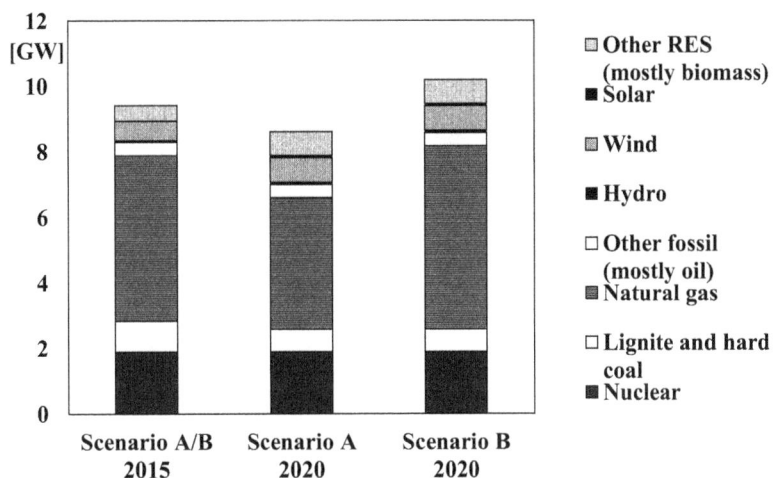

Figure 2. SO&AF scenarios for 2015 and 2020

Basic assumptions for the scenarios

Concerning RES generating capacity, both Scenario A and B assess the expected power balances adjusted to the National Renewable Action Plans submitted by Hungary to the European Commission, as required by Article 4 of Directive 2009/28/EC (Table 4). This document is indicative only for Hungary, since it is subject to revision according to Government Decree 1491/2012. (XI.13.).

For the prospective analysis of thermal and nuclear generating capacity, the existing power plant portfolio, the expected decommissioning of power plants and possible new capacity additions should be assessed. In the SO&AF scenarios, the existing power plant portfolio at the end of 2012 was taken as a basis. The decommissioning of large power plant units was assessed individually, based on the technical lifetime of power generating assets indicated in the licences issued by the regulatory authority. For smaller power plants, a decommissioning rate was estimated. For large-scale new capacity additions, the information on new projects included in the 2012 generation adequacy report prepared by the transmission system operator was used.

Table 4. Projected increase of RES generating capacity according to the NREAP of Hungary

	2010 [MW]	2015 [MW]	2020 [MW]
Hydro	51	52	55
Geothermal	0	4	57
Solar	0	19	63
Wind	330	577	750
Solid	360	377	500
Biogas	13	43	100

Table 5 contains the total net generating capacity and number of two-state units assumed for each category. Smaller thermal power plants were modelled separately in an aggregated way, as described in the methodology; therefore their net generating capacity is not covered in Table 5. Units already mothballed were also not taken into account in the table. (As a result of unfavourable market conditions, several units with a lower thermal efficiency were mothballed recently, reaching a total unavailable capacity of 1.8 GW by the end of 2012.)

Table 5. Net generating capacity and number of power plant units represented by two-state model

	SO&AF 2015 Scenario A and B		SO&AF 2020 Scenario A		SO&AF 2020 Scenario B	
	[MW]	units	[MW]	units	[MW]	units
Nuclear	1,892	4	1,892	4	1,892	4
Lignite	850	5	680	3	680	3
Hard coal	90	3	0	0	0	0
Natural gas	2,816	17	2,159	16	3,498	19
Oil	407	3	407	3	407	3
Other	80	2	130	3	80	2
Total	6,135	34	5,268	29	6,557	31

As described earlier, Scenarios A and B are identical for 2015, since these forecasts look only one year ahead that is negligible compared to the longer lead time of large thermal and nuclear investments. However, the thermal generating assumptions for the 2020 scenarios differ significantly in gas fired net generating capacity.

Scenario A reflects a pessimistic approach where already existing units that are expected to remain in operation until 2020 are only considered. Scenario B as a best estimate projection takes into account three further CCGT units with a total of 1,339 MW of net generating capacity. Various new gas fired CCGT projects were developed in the recent years including both green-field investment and CCGT-upgrade of existing conventional units, but final investment decisions are delayed due to unfavourable market conditions. It was assumed that three of these projects can be realised until 2020 in order to ensure a satisfactory level of national generation adequacy.

The assumptions for nuclear generating capacity are the same, since the existing units of Paks nuclear power station are expected to be decommissioned after 2030. Two units of the lignite fired Mátra power station can be retired before 2020 (reflected in both Scenario A and B), while no new lignite based project is considered at the moment.

Results of probabilistic generation adequacy assessment

Probabilistic generation adequacy analysis was done for the years 2015 (mid-term) and 2020 (long-term), both for Scenario A and B. A one-hour resolution model was used, containing hourly system load, average production of hydro power plants, and hourly aggregated production profile of solar PV generators, small-scale RES and non-RES power plants. For wind production and large power plant units, the COPT values were calculated on a monthly basis, considering maintenance and partially unavailable capacity due to seasonal effects, as well.

Table 6 compares the COPT calculation results for 2015 (Scenarios A/B) and 2020 (Scenarios A and B). Since the calculation was done on a monthly basis, only the probability values computed for December are shown in the table. As described earlier in the methodology, COPT values express the cumulative probability $P(X \geq x)$ of having a system capacity outage greater than or equal to x.

The probabilities of having a certain level of system capacity outage are determined by the total net generating capacity and its composition (share of variable generation, individual forced outage rates of the units). As shown in Figure 2, Scenario B, year 2020 has the highest net generating capacity, while Scenario A, year 2020 has the lowest installed capacity. The level of variable generation differs slightly: 580 MW (Scenario A/B 2015) and 750 MW (Scenario A/B 2020). As a result of all these influencing factors, the COPT values computed for Scenario B 2020 (largest net generating capacity) are the highest, while the probabilities for Scenario A/B 2015 (lower level of wind generating capacity) are the lowest.

Table 6. Capacity outage probability table

Capacity outage [MW]	Scenario A/B 2015 probability $P(X \geq x)$	Scenario A 2020 probability $P(X \geq x)$	Scenario B 2020 probability $P(X \geq x)$
0	1	1	1
250	0.94702914	0.95807864	0.96291066
500	0.72315155	0.82406386	0.84385956
750	0.30692813	0.47417549	0.52752663
1,000	0.11758156	0.16099380	0.23142160
1,250	0.03350453	0.04680618	0.08121181
1,500	0.00828225	0.01231138	0.02844538
1,750	0.00178860	0.00243082	0.00711312
2,000	0.00032328	0.00042170	0.00172182

Based on the COPT data, annual LOLE (h) values and LOLP indices at the annual peak load were calculated, as shown in Table 7. (The LOLP indices are given for the hour of the annual peak load).

Table 7. LOLE (h) values and LOLP at the annual peak load

Year/Scenario	LOLE (h)	LOLP
SO&AF 2015, Scenario A and B	13.54	1.87%
SO&AF 2020, Scenario A	844.12	91.73%
SO&AF 2020, Scenario B	10.82	0.70%

The reliability indices can help to show to what extent the power system is supposed to rely on imported electricity in the three cases that were examined. Based on the adequacy indicators, a slight shortage of domestic generating capacity is expected by 2015, since the LOLP at the annual peak load (without import) exceeds the reference value of 1%. However, more serious electricity import dependency is foreseen for 2020 if no domestic investments will take place. In case of the conservative capacity forecast Scenario A, both LOLE (h) and LOLP values for the year 2020 are extremely high. The LOLP value of near 100% suggests that the load is very likely to exceed the domestically available generating capacity at the annual peak time. Depending on the future evolution of demand and supply side, a close monitoring of the availability of imported electricity might be necessary.

CONCLUSION

Many external factors including economic, regulatory and policy framework affect mid- and long term generation adequacy. Despite of the limited opportunities to consider all these influencing factors, adequacy assessment models and adequacy indicators can provide essential information on possible future capacity shortages for decision-makers. The presented model was developed to enable country level mid- and long term generation adequacy analysis and the computation of related generation adequacy indicators. When applying the methodology to the Hungarian power system, it enables to take into account the capacity credit of the developing wind energy sources more properly. Two projections were analysed covering the years 2015 and 2020. Based on the adequacy indicators, a minor shortage of domestic generating capacity can be expected by 2015. For 2020, a more serious import dependency is foreseen in the absence of investments in power generation.

As stated by the IEEE Task Force on the Capacity Value of Wind Power, the multistate model constructed from a histogram of the wind power output can be considered as an approximate methodology [1], since important factors like the seasonal and diurnal patterns of wind production, and the information on wind/load correlation are not addressed by the model. Despite these concerns, the use of this approach can be justified in the initial phase of RES development when a limited amount of data is available for probabilistic generation adequacy assessment.

NOMENCLATURE

P	Individual probability of capacity outage states for multi-state units	[-]
x	Capacity outage	[MW]
C	Net generating capacity (two-state units) or capacity outage state (multi-state units)	[MW]

Abbreviations

CCGT	Combined Cycle Gas Turbine	[-]
CHP	Combined Heat and Power	[-]
COPT	Capacity Outage Probability Table	[-]
ENTSO-E	European Network of Transmission System Operators for Electricity	[-]
EUE	Expected Unserved Energy	[MWh/year]
EENS	Expected Energy Not Supplied	[MWh/year]
FOR	Forced Outage Rate	[-]
LOEE	Loss of Energy Expectation	[MWh/year]
LOLE	Loss of Load Expectation	[h/year, also: d/year]
LOLP	Loss of Load Probability	[-]
SO&AF	Scenario Outlook and Adequacy Forecast	[-]
UCTE	Union for the Coordination of the Transmission of Electricity	[-]
X	Random variable representing the possible capacity outage states of the system	[-]

REFERENCES

1. Keane, A., Milligan, M., Dent, Ch. J., Hasche, B., D'Annunzio, C., Dragoon, K., Holttinen, H., Samaan, N., Söder, L. and O'Malley, M., Capacity Value of Wind Power, *IEEE Transactions on Power Systems,* Vol. 26, No. 2, pp 564-572, 2011.

2. CEER, Assessment of electricity generation adequacy in European Countries, http://www.ceer.eu/portal/page/portal/EER_HOME/EER_PUBLICATIONS/CEER_P

APERS/Electricity/Tab3/C13-EES-32-03_Generation   Adequacy   Assessment Elec_10-Dec-2013.pdf, [Accessed: 06-March-2014]

3. Allan, R. and Billinton, R., Probabilistic Assessment of Power Systems, *Proceedings of the IEEE,* Vol. 88, No. 2, 2000.

4. Billinton, R. and Bai, G., Generating Capacity Adequacy Associated with Wind Energy, *IEEE Transactions on Energy Conversion,* Vol. 19, No. 3, pp 641-646, 2004.

5. Karki, R., Hu, P. and Billinton, R., A Simplified Wind Power Generation Model for Reliability evaluation, *IEEE Transactions on Energy Conversion,* Vol. 21, No. 2, pp 533-540, 2006.

6. Billinton, R. and Gao, Y., Multistate Wind Energy Conversion System Models for adequacy assessment of Generating Systems incorporating Wind Energy, *IEEE Transactions on Energy Conversion*, Vol. 23, No. 1, pp 163-170, 2008.

7. D'Annunzio, C. and Santoso, S., Noniterative Method to approximate the Effective Load Carrying Capability of a Wind Plant, *IEEE Transactions on Energy Conversion*, Vol. 23, No. 2, pp 544-550, 2008.

8. D'Annunzio, C., Generation adequacy assessment of Power Systems with Significant Wind Generation, *Ph.D Thesis,* The University of Texas at Austin, Austin, 2009.

9. Amelin, M., Comparison of Capacity Credit Calculation Methods for Conventional Power Plants and Wind Power, *IEEE Transactions on Power Systems*, Vol. 24, No. 2, pp 685-691, 2009.

10. Radics, K. and Bartholy, J., Estimating and modelling the Wind Resource of Hungary, *Renewable and Sustainable Energy Reviews*, Vol. 12, pp 874-882, 2008.

11. ENTSO-E, Country Data Packages, https://www.entsoe.eu/data/data-portal/country-packages/, [Accessed: 06-March-2014]

12. ENTSO-E, System Adequacy Retrospect, https://www.entsoe.eu/publications/adequacy-retrospect/, [Accessed: 06-March-2014]

13. ENTSO-E, *Scenario Outlook and Adequacy Forecast 2013-2030*, Brussels, Belgium, 2013.
https://www.entsoe.eu/about-entso-e/system-development/system-adequacy-and-market-modeling/soaf-2013-2030/, [Accessed: 06-March-2014]

9

Operation Strategy for a Power Grid Supplied by 100% Renewable Energy at a Cold Region in Japan

Jorge Morel[*], *Shin'ya Obara, Yuta Morizane*

Department of Electrical and Electronic Engineering, Kitami Institute of Technology, Japan

e-mail: jmorel@mail.kitami-it.ac.jp

ABSTRACT

This paper presents an operation strategy for a power system supplied from 100% renewable energy generation in Kitami City, a cold region in Japan. The main goal of this work is the complete elimination of the CO_2 emissions of the city while keeping the power frequency within prescribed limits. Currently, the main energy related issue in Japan is the reduction of CO_2 emissions without depending on nuclear generation. Also, there is a need for the adoption of distributed generation architecture in order to permit local autonomous operation of the system by the local generation of power. As a solution, this paper proposes a strategy to eliminate CO_2 emissions that considers digital simulations using past hourly meteorological data and demand for one year. Results shows that Kitami City can be supplied entirely by renewable generation, reducing its CO_2 emission to zero while keeping the quality of its power grid frequency within permitted limits.

KEYWORDS

Renewable energy, Wind power, Solar power, Tidal power, Storage system, Frequency control, Smart grid.

INTRODUCTION

Concerns about the adverse effects of global warming on the environment and the need of sustainable use of energy resources have been driving governments around the world in the definition of policies regarding the reduction of CO_2 emissions through the utilization of renewable energy sources. The USA and the European Union have been leading the development of clean and sustainable energy technologies by the implementation of specific policies and targets [1, 2].

One of the technologies applied in power engineering that will allow the interconnection of large numbers of intermittent renewable sources, such as wind and solar, is the smart grid. Besides the possibility of interconnection of variable output renewable generators, smart grids offer other benefits such as active participation of customers in the control of the peak demand of the entire system as well as increased customer participation in the electricity market [3].

Japan, due to its strong and reliable power system, has been focusing mainly on nuclear power to achieve the CO_2 emission targets. The Fukushima Daiichi nuclear disaster revealed the real limitations of the system. Furthermore, forced by a strong public opinion resisting nuclear power generation, Japan now has the challenge to rebuild and adapt its power system based mainly on the deployment of intermittent and clean

---

[*] Corresponding author

renewable energy generation, with a less centralized architecture to increase its reliability in the case of natural disasters.

Tohoku Fukushi University's experimental microgrid in Sendai City demonstrated how important microgrids can be for a country frequently menaced by natural disasters. The microgrid was directly affected by the disaster and it showed resilience [4]. The Japanese government has now established policies to achieve a more reliable and resilient power system [5]. Another important benefit of constructing a self-sufficient microgrid is the reduction of long distance power transfer from centralized power plants, reducing losses and congestion in the transmission lines.

In a cold region, such as Kitami, where there is significant variability in seasonal demand with marked low consumption in summer and high demand in winter, there is a need to shift surplus renewable generation from summer to winter. Storage systems or a coordinated operation with the local power utility should be considered. It is also important to grasp the variability of the total supply considering the type and proportion of each intermittent generation. For example, solar photovoltaic power output changes much faster than a wind turbine power output due to lack of inertia in photovoltaic systems. Also, a tidal farm power output is lower and much more predictable in the long and medium term, comparing to that of a wind farm.

Development of energy systems, that considers a range of infrastructure, such as transport, heat and electricity, has been studied for an optimal design in a given region to create a sustainable energy supply system, reduce CO_2 emissions and utilize intermittent renewable generation. Among the leading research groups in this area is Aalborg University in Denmark that has developed an energy system analysis tool called EnergyPLAN and performed the design of energy systems for specific regions in Europe [6-9].

Japanese industry and academia, due to the reasons previously addressed (reliable power grid and focus on nuclear power) have shown relatively slower development on smart or microgrid technologies before the nuclear disaster in 2011. However, a major effort has been put in from the beginning by certain research groups [4, 10, 11]. After the nuclear disaster of 2011, research activities on independent microgrids for local generation and local consumption, containing sustainable renewable energy generation such as wind, solar and tidal power have increased considerably. Operation of microgrids, aiming at the reduction of CO_2 emissions and the safety of energy supply, in cold, urban and remote areas has been studied [12-14].

The more rapid oscillations of renewable generator outputs in a microgrid makes the utilization of fast acting batteries necessary to match instantaneous imbalances, in contrast with a traditional power grid where the outputs of the generation units can be controlled and match the slower-changing aggregated demand. The utilization of sodium-sulfur (NAS) batteries was mainly analysed in the past for suppressing instantaneous or fast changing impacts in the grid, including the possibility of independent active and reactive power control in this type of storage systems [15].

On a longer time frame, more availability of resources during certain seasons and low demand in the same seasons lead to surplus energy that may be stored for future use when needed. For this case, storage systems have been analysed for economic benefit and environmental impacts [16, 17]. In Reference [16], NAS batteries are compared to a storage system based on organic chemical hydride but the dynamic performance (fast charging/discharging capability) of the NAS battery was not analysed. In [17], a system with no storage system is analysed. Here, good CO_2 reductions are obtained despite the absence of battery storage. However, for a 100% renewable supply, batteries may be necessary to shift energy between seasons or to keep frequency balance, as well as to compensate for any transient faults in the system.

For a complete analysis of a power grid, consideration of dynamic properties of supply and demand is of vital importance. All the energy system designs mentioned above do not consider this aspect.

This work demonstrates that a medium size city located in a cold region, with particular annual load characteristics, can be supplied entirely by renewable energy, reducing completely its CO_2 emission without negatively affecting the electricity quality of the system, and completely breaking its dependence on nuclear power generation.

SYSTEM UNDER STUDY

Location

Kitami City is located in a cold region, on the Hokkaido Island in the northern part of Japan, as shown in Figure 1. Kitami has an annual demand characteristic with a high heat-to-power demand ratio during winter. The temperature reaches a minimum of -20 °C in winter and a maximum of 35 °C in summer. Despite the low temperature in winter, the city gets little rainfall and snowfall. The city has rich natural resources such as wind, solar and tidal power that can be utilized for the generation of clean electrical energy. It is one of the richest areas in solar irradiation in Japan. Also, it has open areas with good average wind speeds which can also be exploited for the generation of electricity. Furthermore, the Saroma Lake tidal current speeds offer the possibility of exploitation of tidal generation [18].

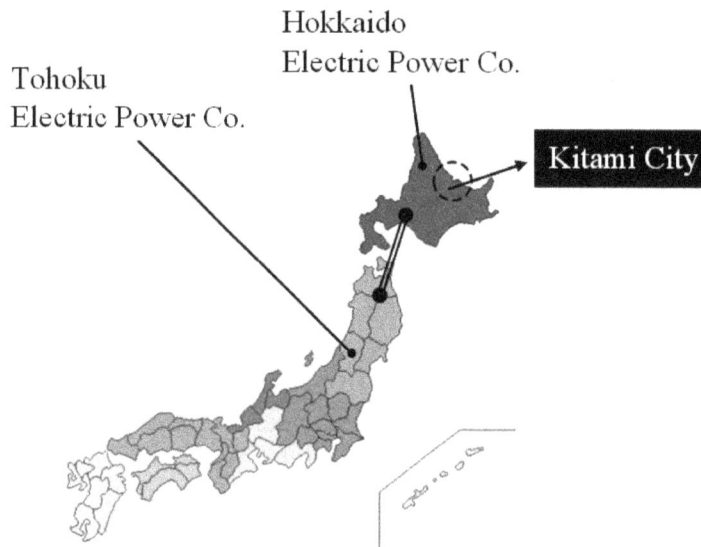

Figure 1. Kitami City location

Transmission network

The Japanese power system consists of ten power companies that supply energy to specific and semi-independent regions. They are interconnected (except Okinawa Electric Power Company) through transmission lines with limited capacities. The Hokkaido Electric Power Company (HEPCO), shown in Figure 1, with a total installed capacity of 7,500 MW, supplies power to the Hokkaido Island, where Kitami City is located. HEPCO is connected to Tohoku Electric Power Company, located in Honshu, the main island of Japan, by a High-Voltage Direct Current (HVDC) transmission system (indicated by a double line in Figure 1), with a capacity of 600 MW, approximately 8% of HEPCO's total installed capacity [19].

The power system of Kitami City is connected to the local utility HEPCO which currently provides the power for the entire city.

The simplified scheme of the Kitami's power system considered for simulation, including the proposed location of renewable generators and storage systems is depicted in Figure 2. The names and rated capacities of the substations are shown in Table 1.

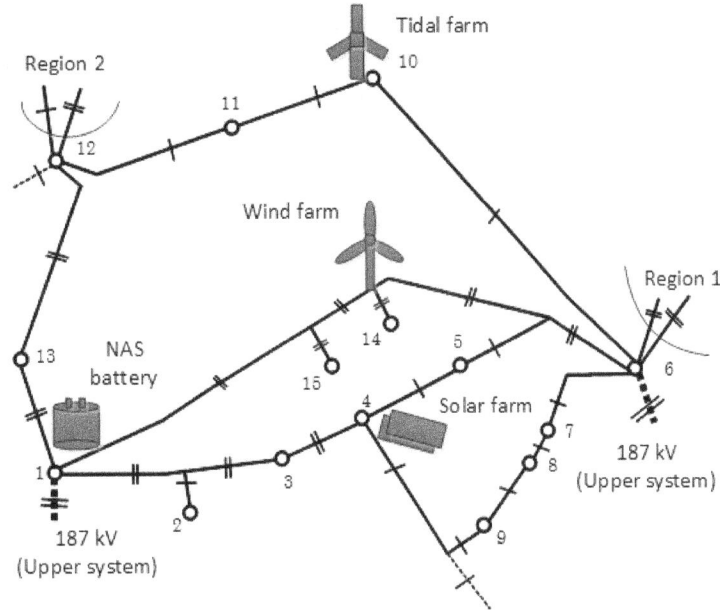

Figure 2. Power system of Kitami City

Selection of the location of the renewable generators was made based on the available resources in the area. Location of the NAS batteries was selected to be the Rubeshibe substation where the bulk power is coming from the conventional power plants of HEPCO. This selection has no direct effect on the results of frequency quality presented in this paper because of the short distances involved. However, from an economic point of view, the most appropriate locations and sizes must be carefully considered.

Table 1. Substation rated capacities

No.	Name	Capacity (MVA)
1	Rubeshibe	280
2	Kuneppu	6
3	Kitaminishi	20
4	Kitami	35
5	Tabata	22
6	Memanbetsu	200
7	Bihoro	20
8	Inami	10
9	Tsubetsu	12
10	Tokoro	12
11	Saroma	10
12	Engaru	18
13	Ikutahara	3
14	Kiyomi	30
15	Ainonai	12

As shown in Figure 2, there are mainly two points of connection to the local utility: The Rubeshibe and the Memanbetsu substations.

Each of them is supplied by a double-circuit transmission line of 187 kV. Region 1 and Region 2 shown in this figure are two systems with no generating units. The total load for Region 1 is 124 MVA and that for Region 2 is 87 MVA. The two thin dashed lines represent the two weak connections to other systems which are not considered in this work. The small circles represent substations and single or double-circuit transmission lines are indicated by one or two short transverse lines over the lines connecting the substations. All lines are overhead, with rated voltage of 66 kV, and short, with lengths of less than 40 km. Typical tower and conductor data for 66 kV transmission line is considered. Each substation is composed of 66 kV/6.6 kV step-down transformers with the rating indicated in Table 1. The reactive power parts of the loads are assumed to be compensated because they do not affect the frequency analysis.

Renewable generation

Wind power. Currently, most of wind turbines (WTs) in the market are variable-speed types: Doubly-Fed Induction Generator (DFIG) and Full-Scale Converter types, which are capable of independently controlling active and reactive power injected to the system. In this work, the wind farm is simulated using an aggregated model of DFIG-based WTs, modelled by Matlab/Simulink.

Solar power. Photovoltaic type solar farms are considered. They are connected to the transmission network via inverters which have the capability of controlling, independently, the amount of reactive power injected to the network for voltage regulation purposes. For frequency study purposes the faster dynamics are not considered; the solar farm is modelled in this work as a first order system with a short time constant of 10 microseconds.

Tidal power. Horizontal axis tidal turbines are considered [20]. They have similar structure and working principle as the WTs. For short-term frequency studies it can be assumed that the turbines have constant power output. For long-term studies the variability can be forecast with a high degree of accuracy. The tidal farm is simulated using an adapted version of the DFIG-based WT. These assumptions are valid due to the similarity between the wind and tidal generation systems and the time scale considered for simulation.

In this paper, only the highly variable power output of the wind and solar farms are simulated as perturbations to the system due to their higher relative outputs and variability compared to those of the tidal farm.

The parameters of the renewable generators are shown in Table 2.

Table 2. Aggregated renewable generator parameters

Generation	Rated capacity (MVA)	Point of connection (No.)
Wind Farm	150	Kiyomi (14)
Solar Farm	100	Kitami (4)
Tidal Farm	1.5	Tokoro (10)

Storage systems

For long-term energy storage aiming at the seasonal and daily energy shift, a storage system with slow dynamics but with high energy density is utilized. For seasonal energy shifting, an organic chemical hydride type system is employed [16].

For instantaneous and fast demand-supply imbalance compensation NAS batteries are employed due to their fast charge and discharge capability. They have been satisfactorily applied in the levelling of power outputs fluctuations of wind farms. They also can be used for load levelling and load peak shaving. [21, 22].

In this work, a simplified model of the NAS battery is considered. A first order system with a small time constant of 10 microseconds is used to represent the fast dynamics of this type of batteries.

OPERATION STRATEGY

The operation strategy is divided in three parts, according to the time frames involved. The overall operation scheme is shown in Figure 3.

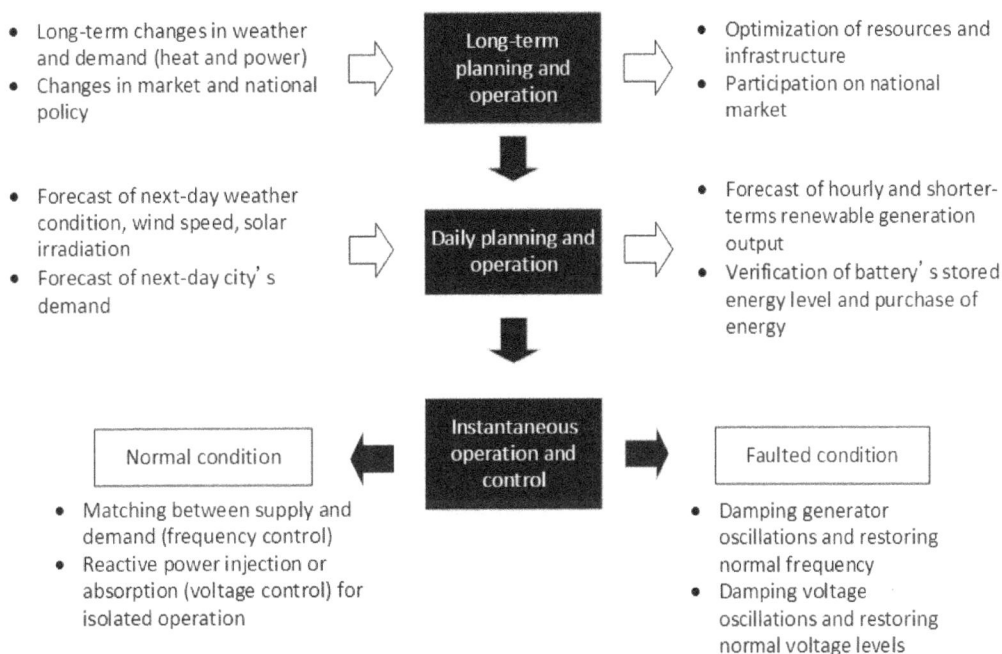

Figure 3. Overall operation scheme

Long-term planning and operation

In order to determine the optimal utilization of the available resources and infrastructure, together with the strategy for the participation of the proposed system in the national electricity market, careful assessment is necessary, focusing on key elements such as weather change, demand trends and change in the energy policy of the entire country .

From the point of view of available and exploitable resources in the city, hourly-averaged balance evaluation between supply and demand, for one year operation, from April 2012 to March 2013, is presented.

The data affecting the output of the renewable generation units is taken from Japanese meteorological and marine agencies, and from the solar power facilities at Kitami Institute of Technology, Japan. The hourly annual energy demand is divided in demand for light and power, and demand for space heating. The power for heating is assumed to be provided by heat pumps.

Daily planning and operation

The demand for the city and the generation output of wind, solar and tidal farms are forecast in advance, for one-day operation. These predictions depend directly on the season, weather condition, tidal current and wind speeds, and solar irradiation.

If the renewable energy to be generated and that stored in the batteries is not enough to supply the forecast demand at any instant in the day the deficit is purchased from the local utility at previously agreed price and time of the day. On the contrary, if there is a surplus of energy, this is sold to the local utility at previously agreed conditions. The net amount of energy interchanged with the local power utility during one year is expected to be equal to zero.

Instantaneous operation and control

The balance between demand and supply is performed instantaneously and automatically by a controller with optimal characteristics since a certain degree of randomness is involved. If the daily forecast of demand and supply is properly established, the last stage in balancing supply and demand is the instantaneous balance of the outputs of generators and energy storages, and the corresponding demand. The instantaneous balancing controls the frequency of the system. Since the renewable generators are highly uncontrollable, with fast changes that may not be compensated by the conventional generators in the system, the control task is performed on the NAS battery for charge and discharge operations.

Furthermore, during faulted conditions, the NAS battery must be capable of dealing with the disturbance by injecting and absorbing active and reactive power to recover the normal operating condition. In case of isolated operation, wind turbines can support frequency by emulating conventional synchronous generators and by injecting reactive power to the system [23, 24]. In this paper, only the normal condition case shown in Figure 3 is considered.

Another important parameter of electricity quality, the voltage level, is assumed to be controlled by the local utility. This assumption is valid since the distances involved are short and the relative size of the target system is small.

For the instantaneous control, during normal operating conditions a Model Predictive Control (MPC) approach is considered due the degree of uncertainty and randomness involved, and due to its capability to deal with multiple inputs and multiple outputs, in contrast to conventional proportional-integral controllers [25].

In Figure 4, the actual frequency of the system f_sys is compared with the reference frequency f_ref of 50 Hz and a signal is sent to the battery for charging and discharging operation (active power, or P control).

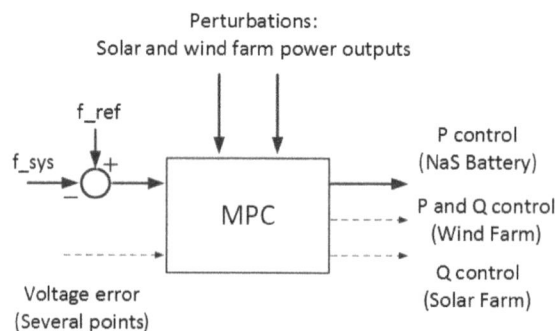

Figure 4. MPC control scheme

The dashed arrows indicate possible inputs and outputs in case of operation of isolated systems where voltage must be entirely controlled by renewable generators (reactive power, or Q control), and the wind turbines can participate in the frequency regulation of the system during abnormal conditions.

SIMULATION RESULTS

First, static simulation is used to assess the total annual renewable generation and the total annual demand in the city, based on actual past data. Secondly, dynamic simulations are performed, to evaluate the impact of the intermittent renewable power outputs in the frequency of the system, and to demonstrate the favourable effects of a fast charging-discharging battery in reducing the frequency oscillations and keeping them within permitted ranges.

Annual supply and demand evaluation

The exploitable renewable power generation together with the annual demand of the city are shown in Figures 5 and 6, respectively. According to results, the total annual demand can be supplied completely by renewable generation. However, shorter-term balances, such as monthly, daily and hourly balances, must be carefully considered in order to define the characteristics of the storage systems needed.

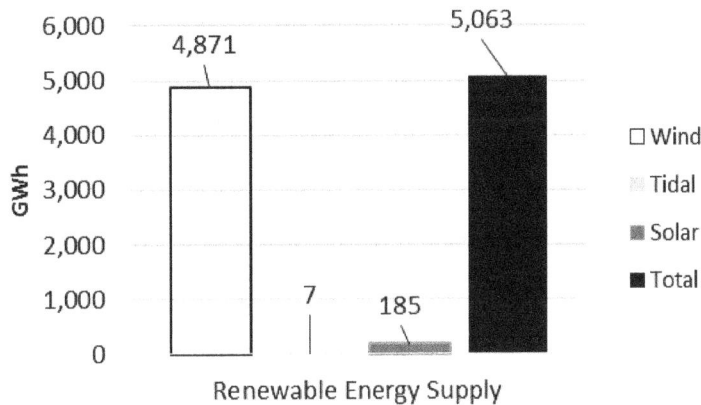

Figure 5. Exploitable renewable energy in Kitami

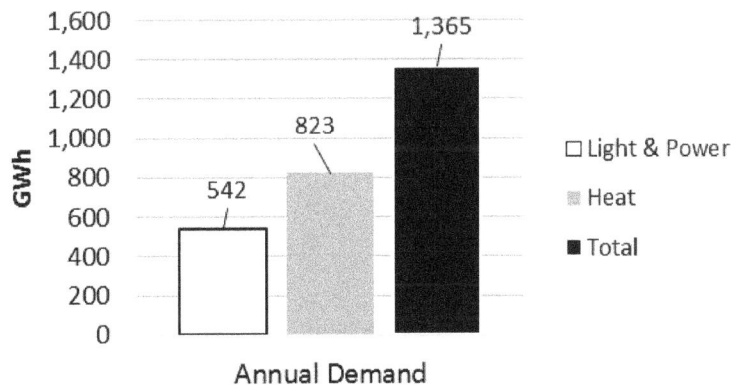

Figure 6. Annual demand of Kitami

Instantaneous energy balance

Three cases are considered. Step change in wind speed, loss of an important load in the system and random variation in solar irradiation.

In this paper, a proportional-integral controller showed similar results to those of the MPC controller because of the single-input single-output characteristic of the cases analysed.

Step change in wind speed. In order to capture the impact of the output of the wind farms, a simple step change in wind speed is selected, keeping other outputs constant. Figure 7 shows the power output of the wind farm due to an increase in wind speed from 10 to 15 m/s, at $t = 20$ s.

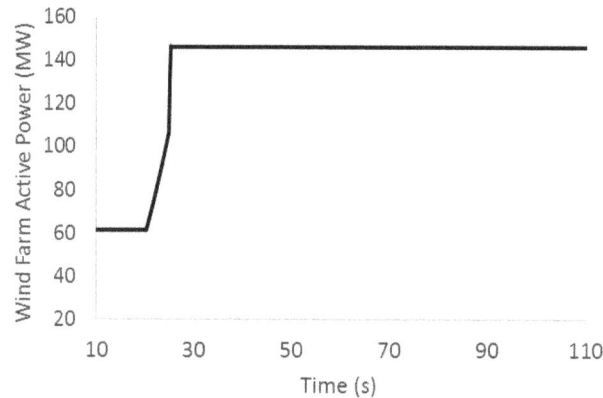

Figure 7. Wind farm power output

It can be noticed that despite the step change in the wind speed the wind turbine output follows a ramp profile. This shows one of the main differences between the characteristics of a wind and a solar farm outputs. Solar farm output follows a much closer profile to the shape in the input due to its lack of inertia.

Figure 8 shows the variations in frequency for the case of wind speed step increase, with and without the action of the controller applied to the NAS battery. Without the support of the fast charging and discharging capability of the battery, the load frequency control (LFC) of the conventional generators is not capable of keeping the frequency within the permitted range of ± 0.3 Hz.

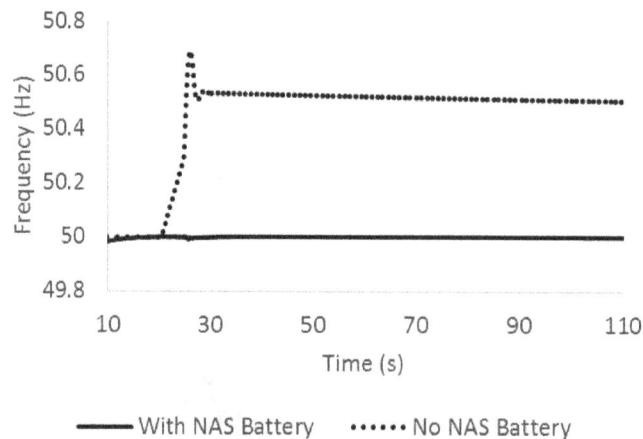

Figure 8. Frequency variation for a step increase in wind speed

Loss of a large load in the system. The loss of a load of 40 MVA at $t = 30$ s is simulated in order to analyse a large instantaneous impact on the system. The loss of a generator or a load behaves as step changes in power that affects the total balance in the system.

Figure 9 shows the variations in frequency for the case of a step change in the system's load, with and without the action of the controller applied to the NAS battery.

Here, as in the case of wind step change, the LFC of conventional generators is not capable of keeping the frequency variation within the range of ± 0.3 Hz. On the other hand, the fast dynamics of the NAS battery allows the variation to remain bounded in this range.

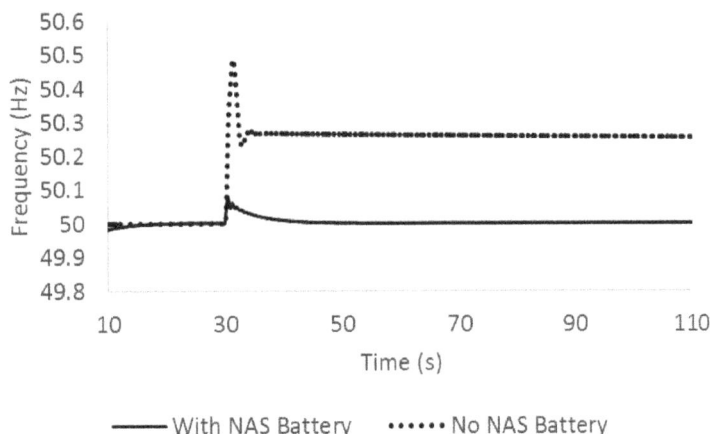

Figure 9. Frequency variation for a loss of load in the system

Figures 10 and 11 show the active power variation of the NAS battery for the two scenarios simulated above. It can be noticed how the NAS battery absorbs satisfactorily the two types of variations in the system, that of wind farm output and that of loss of load.

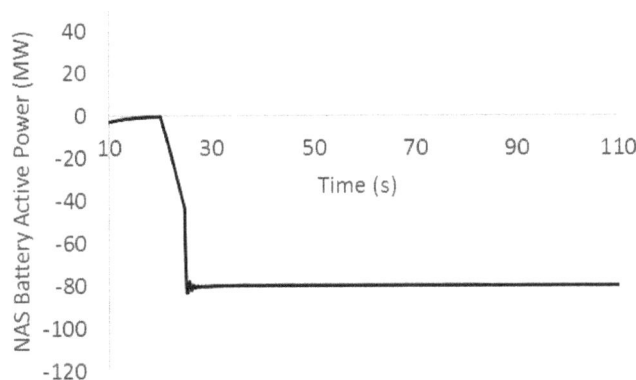

Figure 10. NAS battery active power for a step increase in wind speed

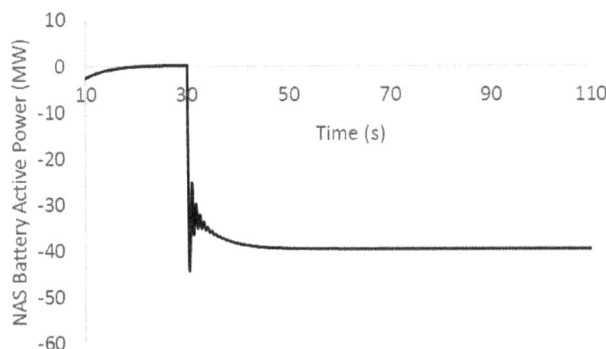

Figure 11. NAS battery active power for a loss of load in the system

Random solar irradiation. A random variation in solar irradiation is considered as the third scenario. The simulations are intended to show how the controller, despite the rapid and random changes, can keep frequency variations within the permitted range of ±0.3 Hz.

Figure 12 shows the power output of the solar farm for random irradiation changes. Due to the fast dynamics of the solar cell, the output follows the random irradiation pattern.

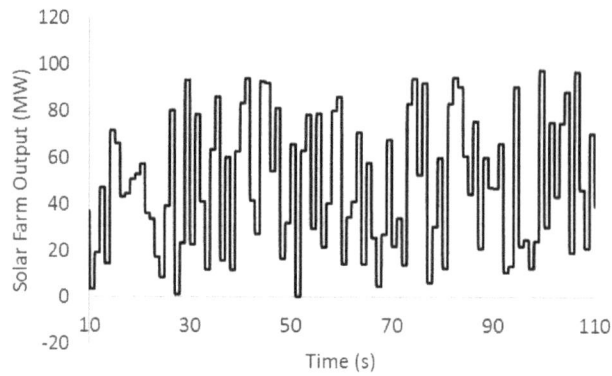

Figure 12. Solar farm power output for random solar irradiation

Figure 13 shows the variation in frequency for the random variation in solar irradiation with and without the action of the controller applied to the NAS battery. It is clear that the LFC of the conventional generators, working without the support of the fast acting NAS battery, is not capable of properly regulating the frequency, due the fast and random changes. The use of a NAS allows the frequency variations to remain within the permitted range.

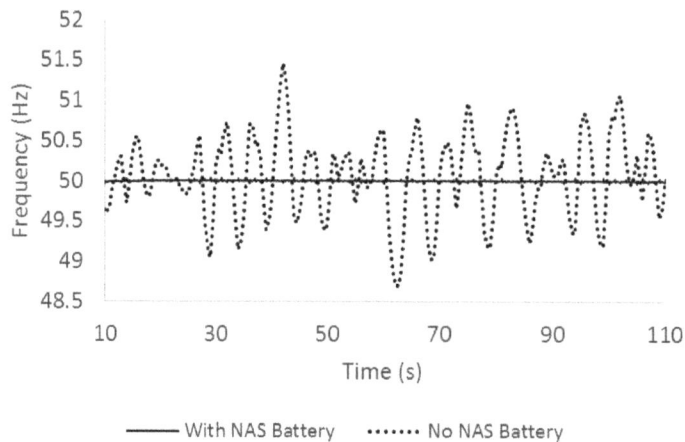

——— With NAS Battery   •••••• No NAS Battery

Figure 13. Frequency variation for random solar irradiation

Figure 14 shows the active power absorption or injection for the case of random variation in solar irradiation. It can be seen how fast the NAS battery change the output set point to follow the changes in solar farm outputs.

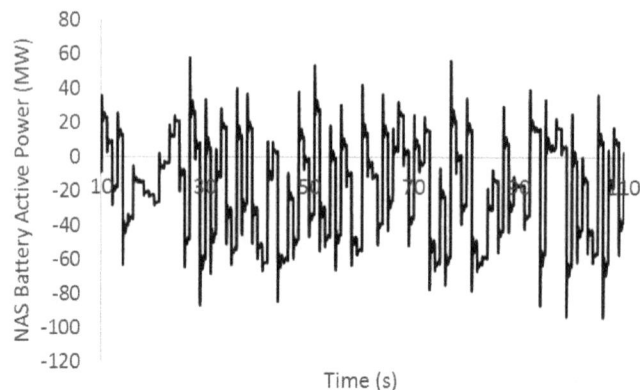

Figure 14. NAS battery active power for random solar irradiation

CONCLUSION

In this paper, a strategy for the operation of renewable energy generators and energy storage devices for the development of a clean smart city, with zero CO_2 emission, and independent from nuclear generation located at a cold region in Japan, is presented. According to simulations based on actual hourly weather and demand data for one year, the city's annual demand can be supplied entirely by solar, wind and tidal generation.

Instantaneous, daily and long-term operational strategies are proposed. The instantaneous strategy that uses a fast dynamic NAS battery and is based on a MPC approach is detailed and dynamically simulated for changes in selected wind speed and solar irradiation, and for a loss of an important load in the system. According to simulations, despite the highly variable outputs of the intermittent sources, the frequency variations can be kept within the permitted range of ± 0.3 Hz by the proposed control system. A MPC-base control system is suggested as it is capable of dealing with discrepancies between the parameters of real systems and that of models utilized for the controller design, as well as randomness, uncertainties and multiple-input multiple-output control in the system.

The area where Kitami City is located is particularly rich in wind, solar and tidal resources. Although the results presented are not directly applicable to an area with less favourable conditions, the results regarding the instantaneous strategy are still applicable as they are more related to short-term power balance.

Since the studied system is considered connected to the local utility's power grid, the voltage control is assumed performed by this stronger system. For an eventual isolated operation scenario, the voltage control must be performed by renewable generators which have converters that are capable of independently controlling active and reactive power. This constitutes a next stage in the development of a completely flexible zero-emission smart city for Kitami.

NOMENCLATURE

f_ref	Frequency Reference	[Hz]
f_sys	System Frequency	[Hz]
P	Active Power	[MW]
Q	Reactive Power	[MW]

Abbreviations

DFIG	Doubly Fed Induction Generator	[-]
HEPCO	Hokkaido Electric Power Company	[-]
LFC	Load Frequency Control	[-]
MPC	Model Predictive Control	[-]
NAS	Sodium-sulphur	[-]
WT	Wind Turbine	[-]

REFERENCES

1. U.S. Department of Energy, Mission, http://www.energy.gov/mission, [Accessed: 24-Feb-2014]
2. European Commission, Energy Strategy for Europe, http://ec.europa.eu/energy/index_en.htm, [Accessed: 24-Febr-2014]
3. SmartGrid.gov, What is a Smart Grid?, https://www.smartgrid.gov/the_smart_grid#smart_grid, [Accessed: 24-Feb-2014]

4. Japan Smart Community Alliance, The Operational Experience of Sendai Microgrid in the Aftermath of the Devastating Earthquake: A Case Study, https://www.smart-japan.org/english/reference/l3/Vcms3_00000020.html, [Accessed: 24-Feb-2014]

5. Ministry of Economy, Trade and Industry (METI), Annual Report on Energy, Outline of the FY2012 Annual Report on Energy (Energy White Paper 2013), http://www.meti.go.jp/english/report/index_whitepaper.html#energy, [Accessed: 24-Feb-2014]

6. EnergyPLAN, Advanced Energy System Analysis Computer Model, Aalborg University, http://www.energyplan.eu/about/, [Accessed: 24-Feb-2014]

7. Lund, H., Andersen, A. N., Østergaard, P. A., Mathiesen, B. V., Conolly, D., From Electricity Smart Grids to Smart Energy Systems - A Market Operation Based Approach and Understanding, *Energy – The International Journal*, Elsevier, Vol. 42, No. 1, pp 96-102, 2012.

8. Conolly, D., Lund, H., Mathiesen, B. V., Werner, S., Möller, B., Persson, U., Boermans, T., Trier, D., Østergaard, P. A., Nielsen, S., Heat Roadmap Europe: Combining District Heating with Heat Savings to Decarbonise the EU Energy System, *Energy Policy*, Elsevier, Vol. 65, pp 475-489, 2014.

9. Conolly, D., Lund, H., Mathiesen, B. V., Leahy, M., The First Step Towards a 100% Renewable Energy-system for Ireland, *Applied Energy*, Elsevier, Vol. 88, No. 2, pp 502-507, 2011.

10. Obara, S., Development of a Dynamic Operational Scheduling Algorithm for an Independent Micro-Grid with Renewable Energy, *Journal of Thermal Science and Technology*, The Japan Society of Mechanical Engineers (JSME), Vol. 3, No. 3, pp 474-485, 2008.

11. U.S. Department of Energy, Microgrids at Berkeley Laboratory, Nagoya 2007, Symposium on Microgrids, Overview of Micro-grid R&D in Japan, http://der.lbl.gov/sites/der.lbl.gov/files/nagoya_morozumi.pdf, [Accessed: 25-February-2014]

12. Obara, S., Kawai, M., Kawae, O., Morizane, Y., Operational Planning of an Independent Microgrid Containing Tidal Power Generators, SOFCs, and Photovoltaics, *Applied Energy*, Elsevier, Vol. 102, pp 1343-1357, 2013.

13. Cellura, M., Di Gangi, A., Orioli, A., Assessment of Energy and Economic Effectiveness of Photovoltaic Systems Operating in a Dense Urban Context, *Journal of Sustainable Development of Energy, Water and Environment Systems*, Issue 1, No. 2, pp 109-121, 2013.

14. Quoilin, S., Orosz, M., Rural Electrification through Decentralized Concentrating Solar Power: Technological and Socio-Economic Aspects, *Journal of Sustainable Development of Energy, Water and Environment Systems,* Issue 1, No. 2, pp 199-212, 2013.

15. Ohtaka, T., Iwamoto, S., A Method for Suppressing Line Overload Phenomena Using NAS Battery Systems, *Electrical Engineering in Japan*, Wiley, Vol. 151, No. 3, pp 19-31, 2005.

16. Obara, S., Morizane, Y., Morel, J., Economic Efficiency of a Renewable Energy Independent Microgrid with Energy Storage by a Sodium-Sulfur Battery or Organic Chemical Hydride, *International Journal of Hydrogen Energy*, Elsevier, Vol. 38, No. 21, pp 8888-8902, 2013.

17. Obara, S., Morel, J., Microgrid Composed of Three or More SOF Combined Cycles without Accumulation of Electricity, *International Journal of Hydrogen Energy*,

Elsevier, Vol. 39, No. 5, pp 2297-2312, 2014.

18. Japan Meteorological Agency, http://www.jma.go.jp/jma/indexe.html, [Accessed: 26-Feb-2014]

19. Hokkaido Electric Power Company, Main Infrastructure, http://www.hepco.co.jp/corporate/ele_power/ele_power.html, [Accessed: 26-Feb-2014] (In Japanese)

20. Tocardo Tidal Turbines, http://www.tocardo.com/, [Accessed: 26-Feb-2014]

21. Tanaka, K., Kurashima, Y., Tamakoshi, T., Recent Sodium Sulfur Battery Application in Japan, *Bonneville Power Administration*, http://www.bpa.gov/Energy/n/tech/energyweb/docs/Energy%20Storage/NGK-Paper.PDF, [Accessed: 26-Feb-2014]

22. NGK Insulators Ltd, http://www.ngk.co.jp/english/products/power/nas/, [Accessed: 26-Feb-2014]

23. Mauricio, J. M., Marano, A., Gomez-Exposito A., Martinez Ramos J. L., Frequency Regulation Contribution Through Variable-Speed Wind Energy Conversion Systems, *IEEE Transactions on Power Systems*, Vol. 24, No. 1, pp 173-180, 2009.

24. Morel, J., Bevrani, H., Ishii, T. and Hiyama, T, A Robust Control Approach for Primary Frequency Regulation Through Variable Speed Wind Turbines, *IEEJ Transactions on Power and Energy*, Vol. 130, No. 11, pp 1002-1009, 2010.

25. Bemporad, A., Morarim M., Ricker N. L., Model Predictive Control Toolbox User's Guide, *The MathWorks, Inc.*, 1998.

10

Evaluation of an Organic Waste Composting Device to Household Treatment

Susana Boeykens[*1], *C. Alejandro Falcó*[2], *Maria Macarena Ruiz Vázquez*[3], *María Del Carmen Tortorelli*[4]

[1]Heterogeneous Systems Chemistry Laboratory, Faculty of Engineering, University of Buenos Aires, P. Colon 850 BA, Buenos Aires, Argentina
e-mail: sboeyke@gmail.com

[2]Heterogeneous Systems Chemistry Laboratory, Faculty of Engineering, University of Buenos Aires, P. Colon 850 BA Buenos Aires, Argentina
e-mail: afalco@fundacion-enlaces.org

[3]Heterogeneous Systems Chemistry Laboratory, Faculty of Engineering, University of Buenos Aires, P. Colon 850 BA, Buenos Aires, Argentina
e-mail: maca@gmail.com

[4]Ecotoxicology Research Program, Department of Basic Sciences, National University of Luján, Ruta 5 y Avenida Constitución, Buenos Aires, Argentina
e-mail: mctortorelli@fibertel.com.ar

ABSTRACT

The performance of a plug-flow automated aerobic digester used with the compost of the Biodegradable Organic Waste (BOW) from a typical family at its generation rhythm was evaluated. During a 13 month assessment, 179.7 kg of BOW were treated and 106.7 kg of compost were obtained with a C:N ratio of 12 and an average concentration of N of about 2.72%. Additional tests enabled to assess the generation of stable and good quality compost according to the considered standards, suitable for using as organic fertilizer and other uses, such as biotreatments. The design, location and operational characteristics of the device have determined reduced leachate emissions, the absence of unpleasant odour generation and incidence of insects or other vectors, implying the viability of their use without affecting the user´s life quality. It could be an efficient alternative treatment for household BOW, from a technical, economic, energy, cultural and environmental point of view, easy to implement for users lacking in special training.

KEYWORDS

Automated aerobic digesters, Household waste treatment, Composting, Plug-flow Biodigester.

INTRODUCTION

Municipal Solid Waste (MSW) treatment is one of the most visible environmental issues in urban areas of Argentina. This problem has been increasing as a result of urbanization and industrialization process, changes related with consumption habits of the growing population and their economic development. Urban communities in Argentina have found increasing difficulties associated with the MSW management due to political, regulatory, technical, institutional, financial problems, in addition with a reduced involvement of the citizens. In the Autonomous City of Buenos Aires, each person generates, in average, more than 1.23 kg of waste per day. In 2012 the total waste generation reported by the Ecological Coordination of the Metropolitan Area, Society of

---

* Corresponding author

the State (CEAMSE), which was disposed without previous separation treatments in the Final Disposal Centres, was 6.5 million tons (6,484,229 tons).

At the other end of MSW management, we can choose cases such as Copenhagen, Denmark, where each person generates about 1.3 kg per day, but only 3% of this waste generation is finally disposed in landfills [1]. In Veracruz, México, segregation at source has led to a massive and positive participation of citizens, reaching a recycling rate of 56% of MSW generated, with an incineration of 39% of MSW segregated, which has been used to produce electricity [2].

Among the various strategies that have been carried out in Denmark to successfully support the MSW management, some cities have offered compost devices free of charge to homeowners to compost their own organic food waste, consuming the entire compost generated in the family's garden [3]. This has greatly reduced the amount of waste that must be transported, separated, treated and disposed, with a significant impact on costs for the city. Besides, the food fraction segregation at source significantly has increased the recyclability of other waste streams, decreasing and eliminating costs of washing.

Household composting should not be seen as an alternative treatment option for all organic waste in a region, but as a complementary solution. It provides a flexible and low-cost solution to cooperate with waste management and facilitate sustainable recycling. Nevertheless, it requires the active participation of a significant proportion of citizens in order to have an impact on waste diversion rate. In comparison with community composting, the economic and environmental benefits associated with household composting, involve reduction of costs and impacts of the temporary disposal on the streets, reducing the expense of collection and transportation of organic waste, and leachate emissions in compactor collection trucks and transfer stations. Additionally, odours on the streets are reduced, also improving the epidemiological vectors control such as flies, rats, etc. Another significant advantage of composting both centralized and at home, is that the compost generated can be used as soil conditioner and amendment, replacing the use of synthetic chemicals. It improves the physical properties of the soil, increasing its water retention and essential nutrients contents [4-7].

There are a variety of implementation strategies for the composting process. Each family in each region has its own food culture, so a large dispersion in the characteristics of raw food waste can be generated, as well as in those of the compost obtained. Also, there could be different destinations for the compost that would determine their respective conditioning. It makes a complex system of multiple variables, which undoubtedly requires further and deeper studies in accordance with its potential direct impact on our own life quality [3, 8].

One of the obstacles in achieving a high percentage of residents involved in the implementation process of compost at home is the idea that the operation is "dirty" and it require a lot of personal intervention, added to prejudices about the inexorable existence of odors and flies, and the associated cultural refusal to make contact with "the garbage". An applicable strategy to accomplish active participation of a significant proportion of citizens is to offer low-cost small size devices, sufficiently automatic and easy to use, with features comparable to those of a simple-use appliance.

The objective of this research was to evaluate the efficiency of a device designed and built to operate automatically at home for household composting. The real scale biodigester is in its third year of receiving uninterruptedly, at the rhythm of the generation, all the food waste from a typical middle class family. The design of this new device was focused on the solution to the problems found with other types of bioreactors [7-10]. These authors suggest procedures that involve discontinue the charge during a period of three months to achieve the compost maturation before sampling for analysis.

Consequently, the process was associated with a change of the ideal operating conditions provided by the need of dispose the wastes as they are generated. In this work, a plug-flow operation is proposed: the waste enters through a receiving chamber, and it is removed from the last maturation chamber, at the other end of the device.

METHODOLOGY

Materials used for composting

The qualitative composition of the BOW corresponds to residues of animal and vegetable products generated during the preparation of family meals, from dishes and kitchen utensils cleaning, and other organic residues (such as paper napkins, etc.). Green wastes from garden pruning were also included when the digester received abundant meat scraps.

The study period was extended by thirteen months (February 1, 2011 until March 1, 2012), totalling 394 days. During this time, the digester received a total amount of 179.7 kg of ROB, which equates to an average of 446 g/day. In the case of the family considered, this volume was associated with the generation of 112 g per day per person of BOW, which implies a low rate of waste generation in comparison to average of this indicator. Adequate samples were taken for analysis from removed compost.

At the start of the study, the net weight of organic matter and biota existing within the digester was 57.00 kg, coincident with the weight on the end date of the evaluation to simplify the accumulation variable in the overall mass balance.

Biodigester description

The composting operation was performed with a plug flow aerobic digester device without forced aeration [11]. The device has a cylindrical geometry, with a total internal volume of 0.14 m³ divided into three chambers (Figure 1). It was built of stainless steel with polyurethane insulation. It has an automatic rotation and programmable system (rotation speed range: 0.5-10 rpm), allowing the rotation frequency adjustment with a sensitivity of min/week.

Slow and pulsating flow between the chambers is regulated by a control system, included in the biodigestor mechanism. The controlled variables are speed, frequency and duration of rotation, aperture of the gates dividing plates between cameras, and horizontal inclination angle of the whole.

The fresh BOW is collected at the same rate of its generation in a bucket of 1.5 L. Every time the bucket's content is completed, it is loaded into the digester. The loading is performed through a gate located in the first chamber, which operates with compost in an initial state of mesophilic degradation. In the intermediate chamber, thermally insulated, waste progress to aerobic decomposition stage, associated with the mesophilic phase completion and the thermophilic stage beginning [12]. When the process is finished, the stabilized compost is removed from the gate located in the third chamber, where it is also possible to extract samples for analysis. The opening degree of the slit over the dividing plates controls the waste residence time in each chamber. During the experimentations, it was possible to close one chamber during the time for an evolution study, whereas the other chambers were still operating.

The design includes a filter system located at the bottom of the last chamber, allowing any leaching drip collection on a lower tray.

The absence of forced aeration was foreseen in the design to minimize the energy consumption and to improve the operational autonomy. Aerobic conditions are insured by the device high frequency rotation. Other authors discusses the different types of devices for aerobic household composting that operate at maximum mixing frequencies of 1 rotation/week [9]. For this study, the designed device was scheduled to generate a daily rotation during 1 min, at 3 rpm, equivalent to three rotations per day, resulting in a significant mixing of the content. Power consumption is reduced to the rotation, which is performed by an asynchronous three-phase motor of 0.09 kW nominal power, scheduled to work only 1 min per day for this test. Due to the equipment internal design, waste is crushed during the rotation, using the same energy for mixing and aeration. The device does not consume water or fuel. Due to the low power consumption required, the device could even work autonomously powered by a small solar panel.

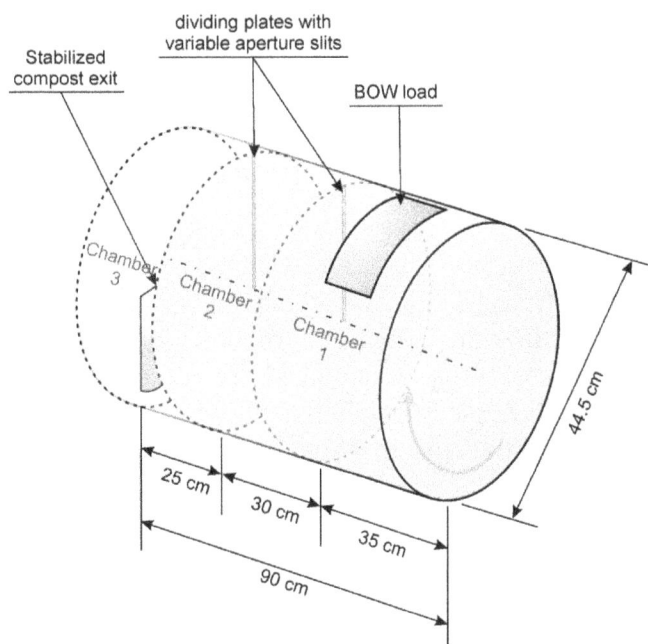

Figure 1. Schematic drawing of the digester with dimensions

Monitoring and analysis

The whole device is mounted on two cells of a Balcoppan Challenger SC103 balance, connected via RS 232 to a computer. Through a specially developed software, the weight is recorded every 15 min automatically. In this way, it is possible to make a correct monitoring of the amount and time when fresh BOW is loaded or stabilized compost is removed, and weight variations experienced by the material mainly because of water exchange with the environment. All the operations are recorded, with a comment, if necessary. The weight of the device body is 23.5 kg (Tare).

As a consequence of the digester content homogeneity, the monitoring sampling was carried out by direct removal. Thus, one sample of each of the chambers was arranged on a plastic cover on the ground, expanded on the surface and divided into four parts, through four proportional diagonal cuts. Of these, two quarters were chosen randomly.

Humidity, temperature, conductivity and pH were weekly studied. The absolute humidity was analysed using a Fischer heater and a DLT-211 Denver Instrument electronic balance [13]. The pH was measured with a pH-222 Lutron portable digital pHmeter and the conductivity was measured with a portable digital conductivity meter Altronix Model CT-2 [14]. The temperature was measured by a manual dial thermometer incorporated into the digester, and a digital portable thermometer was introduced into the compost for a quick reading. Analysis of additional parameters, such as Respirometric Index (RI), Total Organic Carbon (TOC), N, P, Ca, Mg, Na, K, Fe, Zn and Mn concentrations, germination and Root Elongation (RE) tests, and Particle Size Distribution (PSD) were also carried out [14].

For monitoring the presence of insects, special traps were used (Bell Laboratories, Inc.) to determine the type and amount of insects and small mammals circling the digester. They are resistant cardboard traps (approximately 178 mm × 89 mm) with an adhesive portion that holds them to the digester, placed at 3 cm from the gate to BOW entry. Each trap has a highly adhesive surface exposed to the environment where organisms are retained. Four measuring campaigns of insects and small mammals were carried out during the study, using for each opportunity three traps in the digester, with an exposure of five days.

RESULTS AND DISCUSSION

The overall mass balance is:

$$I = O + DbR + A \tag{1}$$

Where I (input) is the mass of BOW load, O (output) is the mass of stabilized compost removed , the term DbR (Disappearance by Reaction) is weight loss that involves the reaction of aerobic composting where organic carbon is gasified to become CO_2, and the term A (accumulation) represents the accumulated mass inside the reactor.

Table 1 summarizes, as example, monthly mass balances during one year of the study. Column 1 represents the sum of the BOW loaded during that month, column 2 the amount of compost removed, column 3 is the difference between the reading of the display of the balance the first day of this month, and the first reading of the following.

Table 1. Monthly and total mass balances

Month	I [kg]	O [kg]	A [kg]
Feb-11	16.8	1.2	0.00
Mar-11	12.24	1.1	9.75
Apr-11	11.6	10.7	-4.45
May-11	16.6	1.7	7.20
Jun-11	12.1	10.2	0.15
Jul-11	10.4	5.3	1.30
Aug-11	14.0	19.4	-7.75
Sep-11	17.0	18.5	-6.25
Oct-11	12.4	11.6	-3.25
Nov-11	19.3	17.3	-4.95
Dec-11	11.0	1.4	3.45
Jan-12	11.5	7.1	-3.00
Feb-12	14.7	1.3	7.80
Average	13.8	8.2	0.00
Total	179.69	106.7	0.00

As a result of the process, there was a weight reduction rate of 40% (wet basis) between the BOW that entered the device and the mature compost generated. Throughout the study, annual average weight reduction values were similar to those indicated. Some previous works examine and compare different composting methods mentioning weight reduction rates (wet basis) in a range of 50-70%, considering this parameter as indicator of the compost stabilization. These values depend on the type of device, BOW, inoculums, and the biotic mixture developed in each digester in their particular conditions. Once the compost stabilization is demonstrated as a result of additional tests, the lower weight reduction means a greater retention of carbon, nitrogen and other nutrients, resulting in better properties for soil amendment [3, 8, 15]. Based on the life cycle analysis of waste treated by household composting, obtaining a stable compost that satisfies the recommended values with the fewest loss of Organic Matter (OM) possible, implies the reduction of the waste carbon footprint, and its impact on climate change.

The residue average residence time on this device was 137 days, which represents the total duration of the BOW composting process for this plug flow system. Figure 2 show the results of a typical weight evolution in two weeks during the study. The general trend in weight reduction inside the device calculated was 0.25% per day. Recorded daily oscillations are due to the exchange of water with the environment, the liquid-vapor equilibrium displacement causes alternatively weight gains and losses of around 2% per day, which means values greater than 1 kg of water were gained and lost throughout the day.

The C:N ratio (total organic carbon/total nitrogen) is used as a compost quality and nitrogen availability indicator [3, 16]. In Argentina, its limit was established in less than 20 for organic amendments, according to the Spanish and Australian regulations. American and Chilean standards, agree with the requirement of a C:N ratio lesser than or equal to 25 for class A compost, and lesser than 30 for class B compost. The rules mentioned above require further analysis to classify into two possible groups. In Group 1, RI values lower than 400 $mg_{O2}/kg_{OM}h$, or self-heating test Dewar below 20 °C are required. In Group 2, RE values above 0.8 are demanded [17-21]. In the case of stabilized compost extracted from the digester under study, the results summarized in Table 2 indicate values to consider that the compost extracted is very mature, it has reached acceptable stabilization to class A, and for both Group 1 or 2.

RE values ≥0.8 indicates no phytotoxic substances presence or very low concentration. A value ≤0.5 may indicate a strong presence of phytotoxic substances and values between 0.5 and 0.8 could be interpreted as moderate presence of these substances. By contrast, if the samples values exceed 1.2, it is considered that the tested product is a growth promoter [20, 25, 26]. Results show evidence of root growth promotion in both species using the compost generated by the device in study.

The results of humidity determinations indicate that they exceed the reference values. The limits considered for moisture in the compost are related to the reduced aeration capacity of the considered systems, and the consequent risks of anaerobiosis and odor generation [20]. In different published works, higher values in the range of 50 to 75% of humidity were found [3, 8, 24]. The ideal humidity content for composting depends on the water retention capacity of the material being composted. In general, high content of OM confers hygroscopicity to the mixture. For the device under study, the daily rotation involves a mechanical contribution to oxygenation allowing higher humidity values and improving the rate of decomposition.

Conductivity is related to the concentration of total soluble, dissolved or suspended salts in the medium. High levels of salinity can be toxic to some sensitive plants. The acceptable level of soluble salts depends on the proposed use of the compost. Results presented here are low enough to very mature compost.

Standards require maximum average compost PSD of 16 mm. Figure 3 shows the particle size distribution of the compost tested, showing compliance with this requirement.

All the campaigns for monitoring presence of insects and small mammals have showed negative results, revealing absence of insects in the digester area. This was confirmed with daily visual observations, accounting absence of insects and unpleasant odor generation.

The volume of leachate generated was approximately 1.1 cm³/kg per day (wet basis). This may explain the high levels of humidity founded in the compost in this work. Other works reported values from zero to 40 cm³/kg per day (wet basis) [3, 10, 27, 28]. The leachate generation is a potential cause of nitrogen loss in stabilized compost and represents environmental impact associated with potential eutrophication as a result of its emission to natural water bodies. Several studies have reported significant loss of nutrients (especially nitrogen) in leachate resulting from composting experiences [8, 29, 30]. It should be noted the high average value of nitrogen registered, significantly higher than the minimum required by the standards. The nitrogen content is reduced during composting, but it is desirable to retain as much as possible [31]. The high value of RE reached evidence the fertilizer capacity of the compost obtained, determining that their use as organic amendment supports crop growth.

Taking into account the characteristics of the compost obtained it is convenient to be used as soil amendment. As a result of the process, 106.7 kg of good quality compost were removed and arranged in the garden and plant pots.

Table 2. Average and reference values of physicochemical and biological variables measured on mature compost removed from the third chamber of the device under study

	Average value	Analytical method	Reference values
Humidity [%]	56.4	SM 2540B	between 30 to 45[21,22] 50 to 70[24]
Density [kg/m³]	652	TMECC 3.01-A	<700[21]
pH	7.2	TMECC 4.11-A	5.0 to 8.5[20,21,22]
TOC [%]	33.9	TMECC 4.02-D	>20[21]
Total N [%]	2.72	TMECC 4.02-D	>0.5[21]
Conductivity [dS/m]	3.2	TMECC 4.10-A	<6[23] <3 Class A,<8 Class B[21] <4[17]
C/N	12	-	<25 Class A, <30 Class B[21,22] <20[17]
P [%]	0.32	TMECC 4.03-A	--
RI [$mg_{O2}/Kg_{OM}h$]	350	TMECC 5.08-B	<400 mg class A[21,22] <500 very mature[20]
RE Lactuca sativa	1.34	TMECC 5.05-A	<0.8 inhibition >1.2 exaltation[20]
RE Raphanus sativum	1.45	TMECC 5.05-A	<0.8 inhibition >1.2 exaltation[20]
Dewar Self-heating test [°C]	1.8	TMECC 5.08-D	<20 °C[20,21] <8 °C[22]
Ca [mg/kg]	54	TMECC 4.05-Ca	--
Mg [mg/kg]	16	TMECC 4.05-Mg	--
Na [mg/kg]	72	TMECC 4.05-Na	--
K [mg/kg]	229	TMECC 4.04-A	--
Fe [mg/kg]	1,352	TMECC 4.05-Fe	--
Zn [mg/kg]	212	TMECC 4.06-Zn	1100 ppm[23]
Mn [mg/kg SM]	337	TMECC 4.05-Mn	--

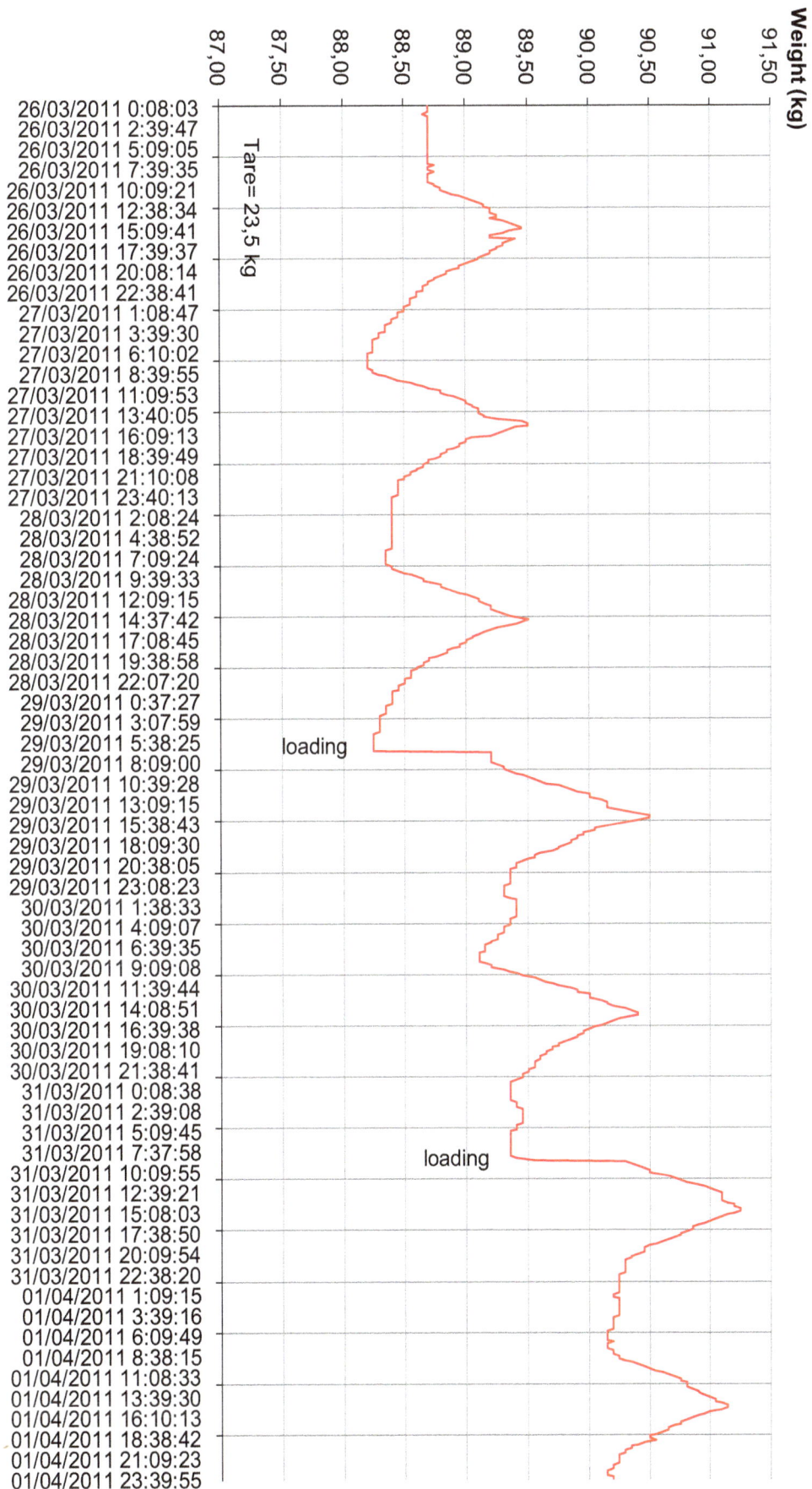

Figure 2. Weight evolution during a typical week

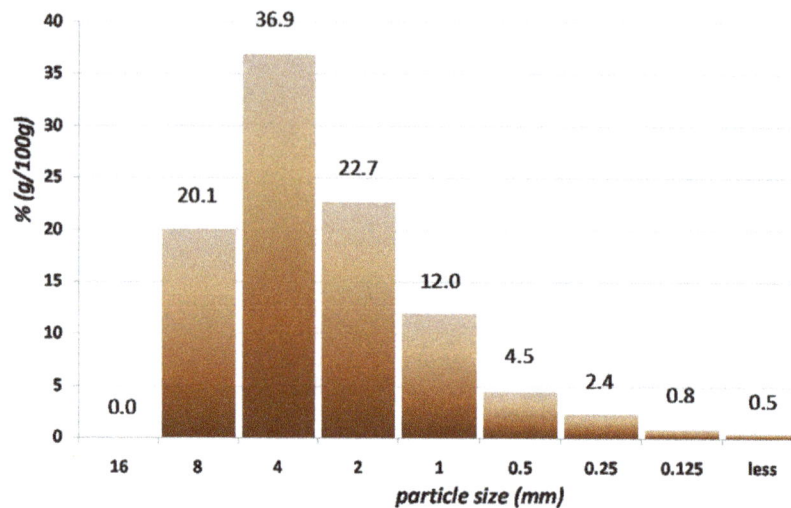

Figure 3. Particle size average distribution of stabilized compost removed from the third chamber of the device under consideration

CONCLUSIONS

According to the reported results, it can be concluded that the compost resulting from BOW treatment on the device studied is a good quality material according to the rules considered, concluding that the tested device has performed a BOW efficient treatment.

The evidence of absence of odor generation and occurrence of insects and small mammals show the viability of installation and operation without impairing the user´s quality of life. Furthermore, device characteristics allow simple and automatic operation, only requiring simple ROB loading and stabilized compost monthly collecting, in both cases through easy gate openings.

Consequently, the results of this study support the conclusion that household composting using the designed device is a simple and suitable alternative treatment for organic waste, from a technical, economic, energy, cultural and environmental point of view, which can be easily implemented by a non-specific trained user, significantly reducing the environmental impact of its own waste.

ACKNOWLEDGEMENTS

This research was supported by the Ministry of Science, Technology and Innovation (Argentina), within the Project: PICT38183/2005, the Buenos Aires University, within the Projects UBACyT I-750/2006, 20020090100102/2010, 20020120100201/2013, and, the NGO: Linkages for Sustainable Development.

REFERENCES

1. Petersen, C., Kielland, M., Statistik for hjemmekompostering 2001 (Statistics for home-composting, in Danish), Miljøprojekt N° 855, Miljøstyrelsen, Miljøministeriet, http://www2.mst.dk/udgiv/publikationer/2003/87-7972-960-6/pdf/87-7972-961-4.pdf, [Accessed: 25-August-2013]
2. Barradas Rebolledo, A., Gestión Integral de Residuos Sólidos Municipales, Estado del Arte, Instituto Tecnológico de Minatitlán, Veracruz, México, 2009.
3. Andersen, J. K., Boldrin, A., Christensen, T. H., Scheutz, C., Mass Balances and Life-cycle Inventory for a Garden Waste Windrow Composting Plant (Aarhus, Denmark), *Waste Manag. Res.* Vol. 31, pp 1934-1942, 2011.

4. McConnell, D. B., Shiralipour, A., Smith, W. H., Agricultural Impact-compost Application improves Soil Properties, *Biocycle* , Vol. 34, pp 61-63, 1993.
5. Jakobsen, S. T., Aerobic Decomposition of Organic Wastes II: Value of Compost as a Fertilizer, *Resour. Conserv. Recycl.*, Vol. 13, pp 57-71, 1995.

6. Hargreaves, J. C., Adl, M. S., Warman, P. R., A Review of the use of Composted Municipal Solid Waste in Agriculture, *Agric. Ecosyst. Environ.*, Vol. 123, pp 1-14, 2008.

7. Martínez-Blanco, J., Muñoz, P., Antón, A., Rieradevall, J., Life Cycle assessment of the use of Compost from Municipal Organic Waste for Fertilization of Tomato Crops, *Resour. Conserv. Recycl.*, Vol. 53, pp 340-51, 2009.

8. Colón, J., Martínez-Blanco, J., Gabarell, X., Artola, A., Sánchez, A., Rieradevall, J., Font, X., Environmental assessment of Home composting, *Resour. Conserv. Recycl.*, Vol. 54, pp 893-904, 2010.
9. Andersen, J. K., Boldrin, A., Christensen, T. H., Scheutz, C., Greenhouse Gas Emissions from Home composting of Organic Household Waste, *Waste Manag.*, Vol. 30, pp 2475-2482, 2010.
10. Amlinger, F., Peyr, S., Cuhls, C., Greenhouse Gas Emissions from composting and Mechanical Biological treatment, *Waste Manag. Res.*, Vol. 26, pp 47-60, 2008.

11. Tchobanoglous, A., Theisen, L., Vigil, S., *Integrated Solid Waste Management*, McGraw-Hill, 1998.
12. Mason, I. G., Mathematical modelling of the Composting Process: A Review, *Waste Manag.*, Vol. 26, pp 3-21, 2006.
13. APHA-AWWA-WEF, *Standard Methods for the Examination of Water and Wastewater*, 22nd Edition, 2012.
14. TMECC, *Test Methods for the Examination of Composting and Compost*, US Composting Council, Bethesda, MD., U.S.A, 2002.
15. Papadopoulos, A. E., Stylianou, M. A., Michalopoulos, C. P., Moustakas, K. G., Hapeshis, K. M., Vogiatzidaki, E. E. I., Loizidou, M. D., Performance of a New Household Composter during In-home testing, *Waste Manag.*, Vol. 29, pp 204-213, 2009.
16. Boldrin, A., Körner, I., Krogmann, U., Christensen, T. H., Composting: Mass Balances and Product Quality, In: Christensen, T. H. (Ed.), *Solid Waste Technology and Management*, John Wiley & Sons Ltd., Chicester, 2010.

17. Servicio Nacional de Sanidad y Calidad Agroalimentaria (SENASA) Resol, N° 264 /2011, Reglamento para el registro de fertilizantes, enmiendas, sustratos, acondicionadores, protectores y materias primas en la República Argentina, Ministerio de Agricultura Ganadería y Pezca, Argentina, 2011.
18. Real Decreto 824/2005, Requerimientos Mínimos de la Calidad de un Compost, Boletín Oficial del Estado, N°171, pp 25592-25669, España, 2005.
19. Australian Standard AS4454, Compost, soil, conditioners and mulches, Standards Australia, 4° Ed., Sydney, NSW, 2012.
20. California Compost Quality Council (CCQC), Compost Maturity Index, Technical Report, 2001.
21. Instituto Nacional de Normalización (INN), Compost-Clasificación y Requisitos, Norma Chilena de Compost, N° 2880/2004 (NCh 2880-2004), Chile, 2004.
22. Canadian Council of Ministers of the Environment (CCME), Guidelines for Compost Quality, Canada, 2005.

23. Council of the European Communities (CEC), Council Directive 1999/31/EC, On the landfill of waste, *Official Journal of the European Communities*, N° L 182/1-19, 1999.

24. Gray, K. R., Sherman, K., Review of composting, Part 1, *Process Biochemistry,* Vol. 6, No. 6, pp 32-36, 1971.

25. Zucconi, F., Pera, A., Forte, M., De Bertoli, M., Evaluating Toxicity in Immature Compost, *Biocycle,* Vol. 22, pp 54-57, 1981.

26. Varnero, M. T., Orellana, R., Rojas, C., Santibáñez, C., Evaluación de especies sensibles a metabolitos fitotóxicos mediante bioensayos de germinación, Tomo III, pp 363-369, in: *El Medioambiente en Iberoamérica: Visión desde la Física y la Química en los albores del Siglo XXI*, Editor: Juan F. Gallardo Lancho, Sociedad Iberoamericana de Física y Química Ambiental, Badajoz, Spain, 2006.

27. Ming, L., Xuya, P., Youcai, Z., Wenchuan, D., Huashuai, C., Guotao, L., Microbial Inoculum with Leachate Recirculated Cultivation for the Enhancement of MSWcomposting, *J. Hazard. Mater.* , Vol. 153, pp 885-91, 2008.

28. Wheeler, P. A., Parfitt, J., Life Cycle assessment of Home composting, *Proceedings of Waste 2002 Conference*, Stratford, UK, 2002.

29. Parkinson, R., Gibbs, P., Burchett, S., Misselbrook, T., Effect of turning Regime and Seasonal Weather Conditions on Nitrogen and Phosphorus losses during Aerobic Composting of Cattle Manure, *Bioresour. Technol.*, Vol. 91, pp 171-8, 2004.

30. Sommer, S. G., Effect of composting on Nutrient loss and Nitrogen availability of Cattle Deep Litter. *Eur. J. Agron.*, Vol. 14, pp 123-33, 2001.

31. Giró, F., El compost como producto, Exigencias en la recogida, estándares y condiciones de aplicación, *Jornada Internacional sobre la Gestión Integrada de Recursos y Residuos en la Unión Europea*, Santander, Spain, 2002.

11

Climate Change and Vulnerabilities of the European Energy Balance

Giuliano Buceti
UTFUS, ENEA, C.R. Frascati, C.P. 65, I-00044, Frascati, Italy
e-mail: giuliano.buceti@enea.it

ABSTRACT

Energy consumption induces climate change but at the same time modifications in climate impact the energy sector both in terms of supply capacity and shift in energy demand. Different regions will be affected in different ways and this paper aims at analysing the issue at the European level. Usually rising sea levels, extremes of weather and an increase in the frequency of droughts and floods are indicated to play havoc with the world's energy systems but they can be hardly estimated and this study will be limited to the effects of the increase in average temperature. Tipping points are also taken out of any quantitative assessment. Structure of the EU energy budget is presented, shifts in energy demand, vulnerabilities of supply and risks for energy infrastructure are discussed in order to eventually provide figures of possible further threats to the continental energy security.

KEYWORDS

Energy security, Climate change, European energy budget, Heating degree days, Cooling degree days, Adaptation, Energy supply, Energy demand.

INTRODUCTION

The relation between energy and climate change is usually debated in terms of how emissions from energy consumption induces alterations in the planet's equilibrium. Indeed, there also exists the issue of how modification of climate impacts the energy sector both in terms of supply capacity and shift in consumption trends.

Assessment of this relation is challenging because reliable forecasts of future climate meet with intrinsic difficulties. First of all, the low predictability of climate as a whole [1] affects the capabilities of making guesses in specific sectors like energy. Second, all forecasts are made under the assumption, reasonable but not certain, that no tipping points [2] will be trespassed. Finally, a shared view among the scholars is that climate will show a double face: the mean, whose effects will be related to the increase of the mean temperature, and the extreme, whose effects come from the increase in frequency and intensity of extreme events. While effects from the increase of the mean temperature are forecasted with a certain degree of confidence, those from extremes are definitely more difficult to predict. Several studies have been carried out on vulnerabilities of energy supply and new trends in consumption both at global [3] and regional [4] level and an extensive US National Climate Assessment [5] has been carried out to specifically address the issue. Nevertheless, because the exact extension of climate change is still indefinite, its effect on the energy sector remains vague. In fact greater uncertainty is on supply and production, affected by extreme weather events, than on energy demand, driven mainly by the increase in mean temperature. Because of this incertitude, most of the literature [6] provides just lists of qualitative trends rather than quantitative evaluations. This paper aims at making a more detailed analysis at European level of the consequences on the EU 27 energy budget and at investigating the exact extent of a possible energy security issue. The analysis is carried out in a +2 °C scenario seen acceptable at the 2010 UNFCCC Cancun Conference even if a recent study suggests this could be accompanied by a significantly changed climate from today, for example in terms of precipitation [7]. In this study will be taken into account only direct effects on energy systems from increase of temperature while indirect effects, like those coming from changes in ecosystems, will be taken out.

STRUCTURE OF EU ENERGY BALANCE

The issue of the energy security, i.e. a possible imbalance between supply and demand, may occur at different time and spatial scales. In this study the focus will be at the level of the annual energy balance. Table 1 shows a condensed version of the EU 27 energy balance for the year 2010 [8] and Figure 1 provides a pictorial view of the same information. The way it is organised is similar to the usual energy balance at national level and contains three sections:

- Supply of primary energy, made by adding up flows of energy entering the continental territory (production and imports) and subtracting flows of energy made unavailable for continental consumption (exports, international bunkers, etc.);
- Transformation + energy industry use + losses, which covers those activities that transform the original primary (and sometimes secondary) commodity into a form which is better suited for specific uses and ready for the final consumption;
- Final energy consumption, obtained by summing up energy spent in industry, residential, services, public administration, transport and so on.

It is worth noting that from 1,759 Mtoe of primary energy, only 1,152 Mtoe become available for final consumption.

The structure could be seen as a matrix whose elements are supply, transformation and consumption vs source types. Climate change impacts each element of the matrix forcing energy needs to evolve. The result will be a new matrix with possible gaps between supply and demand and a potential energy security issues.

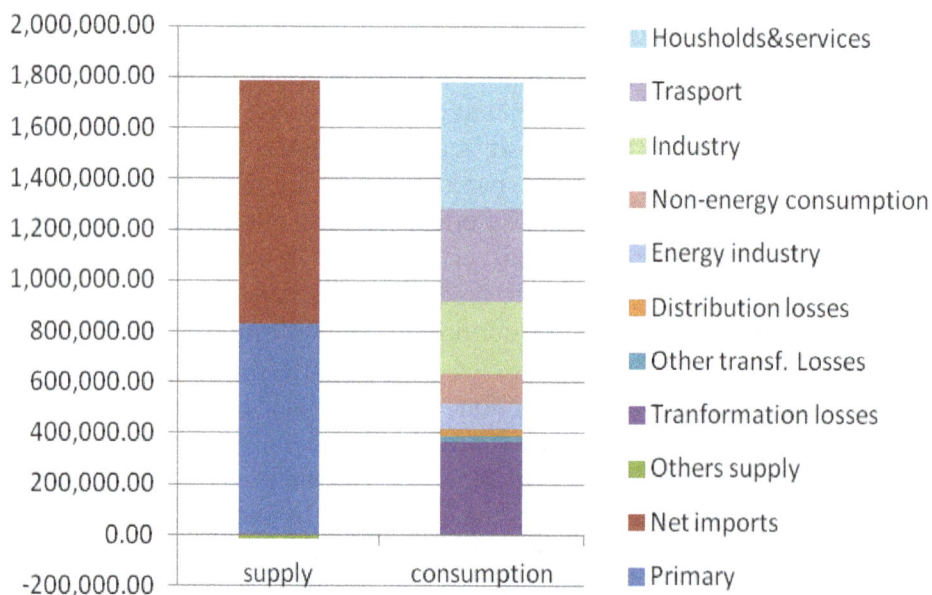

Figure 1. Pictorial view of the EU 27 energy balance year 2010 (Elaboration of Eurostat data)

Table 1. A condensed version of EU 27 energy balance year 2010 (Elaboration of Eurostat data)

EU-27 Year: 2010 1,000 toe	Total all products	All coal	Crude oil and NGL	Feed stocks	Total pet. products	Natural gas	Derived gas	Nuclear heat	Total renew. energy	Other fuels	Derived heat	Electrical energy
Gross inland consumption	1,759,390	2,793.75	6,202.42	11,175	-13,939	441,899	67	236,563	172,326	3,748	-2	299
Primary production	831,105	162,964	94,360	2,997		156,190		236,563	166,851	3,747		
Import-export	951,806											
Other supply	-14,497											
Transformation input	1,400,973	262,176	613,852	52,015	22,366	148,903	8,625	236,563	47,431	1,418	580	165
Transformation output	1,013,230	33,776			655,097		20,730		57		63,562	238,686
Exchanges and transfers. returns	9,444	0	-2,347	40,179	-28,333				-46,432			46,376
Consumption of the energy branch	87,887	880	1		38,825	14,616	3,675		301	0	5,156	24,310
Distribution losses	27,716	45	6		21	4,228	883		25	1	4,494	18,011
Available for final consumption	1,265,488	0	4,036	-661	551,613	274,152	7,614	0	78,194	2,328	53,330	242,876
Final non-energy consumption	114,792	50,046	2,424		97,902	13,146	0		0			
Final energy consumption	1,152,503	55	2,113		454,953	261,170	7,611		78,220	2,328	53,292	242,660
Industry	289,621	48,768	2,113		32,871	84,219	7,553		21,088	2,321	15,457	88,283
Transport	365,117	0	0		343,509	2,435	0		13,328			5,833
Other sectors	497,765	0	0		78,574	174,516	58		43,803	7	37,835	148,544
Households	307,823	13,596	0		42,789	119,334	22		39,750	0	22,661	72,295
Services	152,059	10,461	0		19,472	47,298	36		1,950	7	10,054	71,280
Agriculture/ Forestry	25,068	1,677	0		13,759	3,698	0		1,826	0	290	4,106
Fishing	886	1,388	0		823	1	0		36	0	0	25

SHIFTS IN ENERGY DEMAND

Comprehensive lists of general trends in energy demand driven by climate changes are available in the literature [5]. Certain areas which are expected to play a major role are listed in Table 2.

Table 2. Major driver and trends of changes in energy consumption

Driver	Demand decrease	Demand increase	Sector in energy statistics	Fuel/carrier	Correction factor
Higher mean temperature	Space heating		Residential	Natural gas	f = f(HDD)
		Space cooling	Residential	Electricity	f = f(CDD)
Peak temperature in summer/heat island effects		Electricity peak for space cooling			Local/fine modelling needed
Draught/water scarcity		Irrigation	Agriculture	Electricity	Local/fine modelling needed

Climate change could trigger several drivers of modification. First is the higher mean temperature in itself. Breakdown of EU 27 energy consumption shows that industry, transport and residential sectors take about 90% of the share (see Figure 2). Literature and simulations predict that consumption for transport and industry will be affected little or not at all by increase of temperature. IPCC in 2007 in its Fourth Assessment Report states that: "…*Climate-change vulnerabilities of industry, settlement and society are mainly related to extreme weather events rather than to gradual climate change (very high confidence)…*" [9] and again in 2011: "…*Although the energy, industry, and transportation sectors are of great economic importance, the climate sensitivity of most activities is low relative to that of agriculture and natural ecosystems, while the capacity for autonomous adaptation is high, as long as climate change takes place gradually* [10].

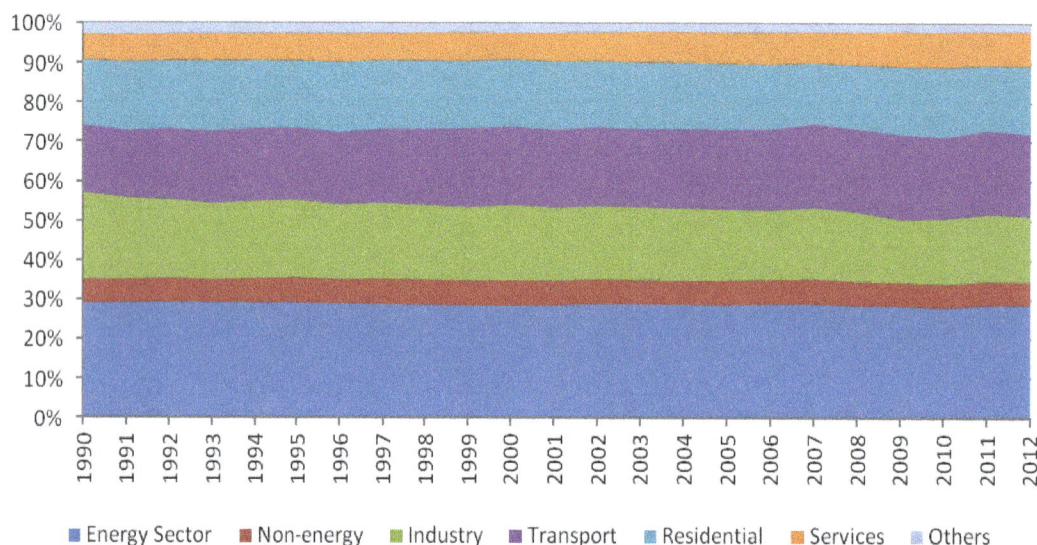

Figure 2. Breakdown of EU 28 energy consumption years 1990-2012 (Elaboration of Eurostat data)

Data on European energy consumption are in agreement with these statements. Figure 3 shows Heating Degree Days (HDD) and energy consumption for industry, transportation and residential from 1990 to 2012 in EU 27. Data are taken from Eurostat database.

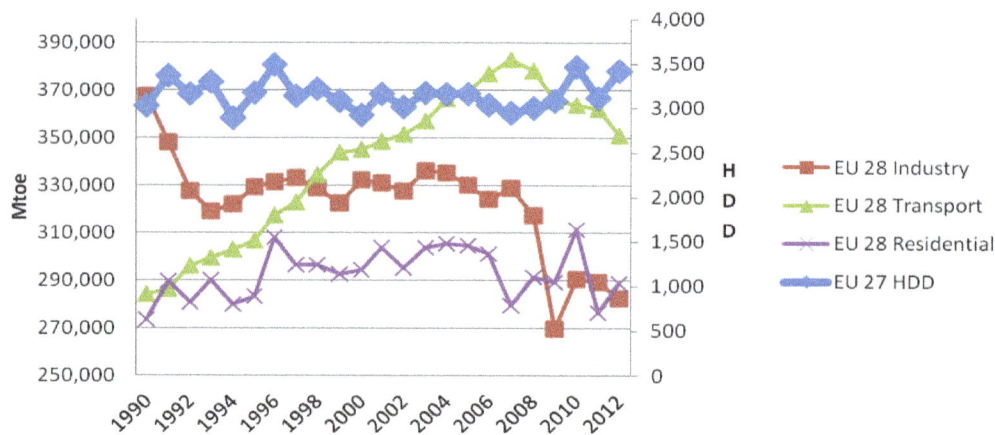

Figure 3. Heating Degree Days (HDD) vs. energy consumption in industry, transportation and residential (households + services) in EU 27, years 1990-2012 (Elaboration of Eurostat data)

At a glance, it appears that industry and transportation are quite independent from HDD while residential sector shows a different behaviour. In fact, calculations give a correlation factor between HDD and industry consumption or HDD vs. transportation is equal to -0.20 while a significant 0.45 between HDD and residential indicates a much stronger relation between climate and consumption in residential sector.

Because industry and transportation sectors represent about 2/3 of energy demand, it appears that the same ratio could be considered climate independent (see Figure 4) and does not entail, from this point of view, issues of energy security. In fact, a warmer world will demand less space heating and more energy for space cooling in summer. Usually these shifts are expressed in HDD and CDD (Cooling Degree Days). In EU-27 in the period 1980-2009 the number of HDD has decreased by 13% [11], yet with substantial inter-annual variation. The pattern shows that the decrease has not been homogeneous across Europe and the absolute decrease has been largest in the cool regions in northern Europe where heating demand is highest. Other studies calculate a 10% reduction of HDD for most locations in Europe [12] under the assumption of a temperature increase of 1 K in winter. In this paper the effect on energy demand for heating is assumed to be in linear relation with HDD and expressed as [13]

$$(\text{Residential \& Services heating})_{acc} = (\text{Residential \& Services heating})_{bcc} \times \_HDD_{acc} / HDD_{bcc} \tag{1}$$

Acc and bcc stand for "after" and "before climate change". The exact value for bcc depends on which year is taken as reference, while acc depends on model, year and selected future scenario. The same reasoning applies to CDD that are expected to increase. Several scholars [14] suggest that the worldwide energy demand for cooling will increase not only to face higher summer temperature but also the increase of cooled surface. Because this is going to happen mostly in non OECD countries, will not be considered in this study on European trends.

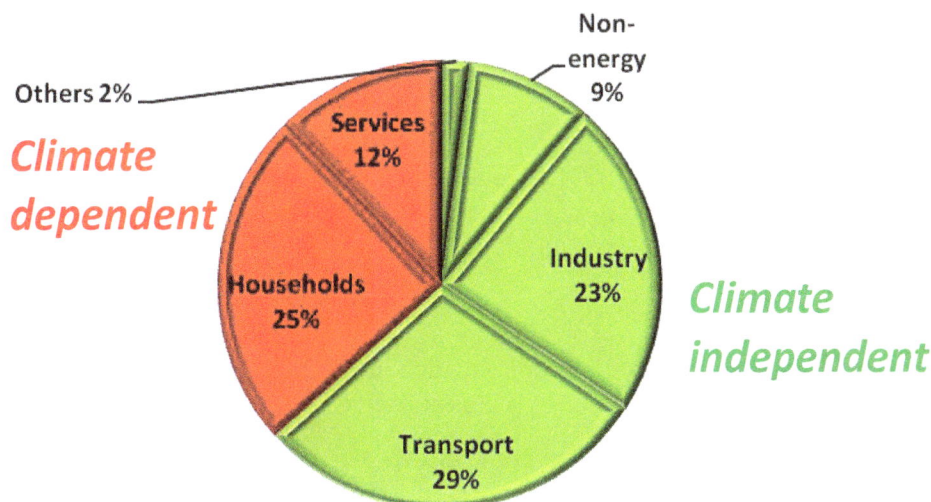

Figure 4. Breakdown of EU27 2010 energy balance in climate sensitive areas (Eurostat data)

Because the EU27 buildings use 23% of the primary energy supply [15] and almost 40% of total final energy consumption [16] is for heating, a possible decrease from 10 to 20% of HDD will imply a saving from 4 to 8% of total energy consumption.

This must be compared with the effect of increase in CDD to have a figure of the net saving. It's useful to remind that surface cooling is done through electric devices and electricity accounts for about 20% of the total EU 27 energy consumption. A breakdown of electricity consumption (see Figure 5) shows that residential + services take together a share of 60% but not more than 5% is used for air conditioning. In Figure 5 this is shown only for services but this applies also to residential [17]. All together the net result is that surface cooling takes about 1% [20% (electricity share of total energy consumption) × 60% (services-residential share) × 5% (air conditioning share)] of total energy consumption. A 65% (Table 3) increase in CDD could even double this figure but in the overall budget, it appears largely offset by the savings from the reduction of HDD. Beside the increase of the average temperature, climate models predict that local peaks in summer temperatures will be much more frequent and more pronounced in absolute value. This does not entail shortage of energy *per se* but, because the total installed power is tailored on peaks of electricity demand, more robust interconnections in and among regions and possible re-sizing of power plant parks will be needed.

Agriculture appears in Table 2 since water scarcity could exacerbate the need of energy for irrigation and increase the total demand in critical summers. In fact, climatic variables, such as temperature and precipitation, are essential inputs to agricultural production and different combinations and seasonal patterns have a direct consequence on yields. That said, agriculture accounts only for about 1% (see Table 1) of total electricity consumption.

In summary, only households and services consumption appear to be climate sensitive producing a combined effect, in a +2 °C scenario, of a possible reduction of 6% in total energy consumption.

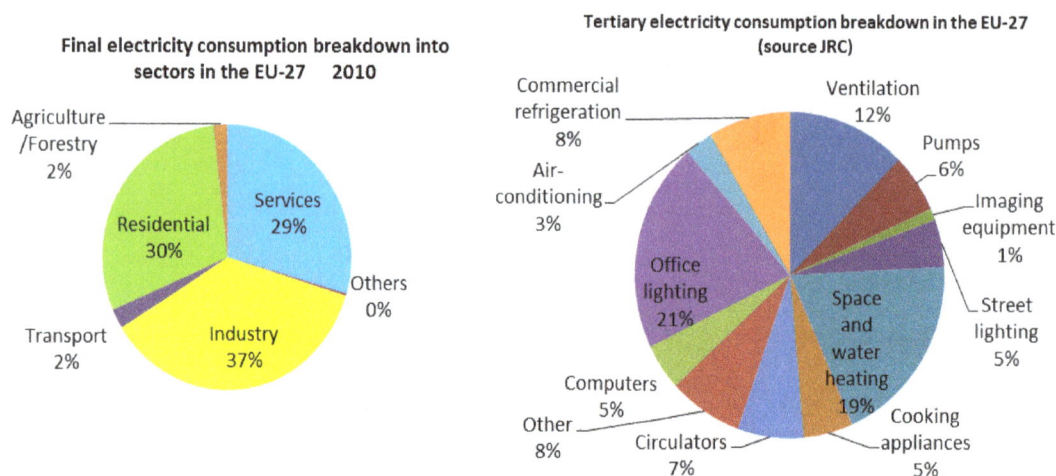

Figure 5. Breakdown of electricity consumption in EU 27, year 2010 (JRC elaboration of Eurostat data)

VULNERABILITIES IN ENERGY SUPPLY

The future of energy supply is much more difficult to predict. Table 4 summarises how this sector could be hit by climate change. It appears that supply based on fossils could be affected in several ways. First in the table are extreme weather events. Type, frequency and intensity of these extremes vary region by region [18] and this raises concerns about energy infrastructures which were built to meet climate conditions of the past and there is no reason to believe that they will meet future conditions. However, there are claims that off-shore gas/oil platforms already today usually operate within extreme weather conditions [19]. Whatever the real resilience is, the only answer to extremes will be adaptation.

This is different from water scarcity, second in the table and most likely a more compelling concern, where proactive initiatives can be put in place in advance. Today, United States and Europe produce 91% and 78% of their total electricity by thermoelectric (nuclear and fossil-fuelled) power plants [20], which directly depend for cooling on the availability and temperature of water resources. This makes the supply of electricity vulnerable to the combined impacts of lower summer river flows and higher river water temperatures. In fact, even when cooling water is available, its temperature is expected to rise [20] and plant efficiency to decline. In particular, for 1 °C increase in air temperature, the power output of natural gas-fired combustion turbines (often used for peaking) is estimated to decrease by approximately 0.6% – 0.7% [21]. In the same condition, nuclear power plants, output losses are estimated to be approximately 0.5% [22] [23]. A further difficulty is that electricity generation, in case of drought, goes in fierce competition with agriculture. This is not listed in Table 3 because it does not directly affect the level of energy supply but could occasionally force to stop or reduce plants operation in case agriculture needs have a higher priority compared with energy production. In order to simplify our scheme, a 0.5% decrease of efficiency per °C will assumed in all thermo/nuclear power plants.

Table 3. Trends in energy supply driven by climate change

Driver	Supply decrease	Supply increase	Sector in energy statistics	Fuel/carrier
Extreme weather events	Oil and gas exploration and production		Oil and gas extraction	Oil and gas
	Disruptions of infrastructures (transport, transmission,...)		All	All
Summer peak temperature/heat island effects/water scarcity/Draught	Thermal power plants stop/reduced operations and lower effiency		Electricity conversion	Fossils and nuclear
	Biomass reduction		Bio production	Biomass
	Reduction of hydropower generation		Hydropower	Hydro
Sea level rise	Threat to coastal power plants and infrastructures		All	All
Cloud and dust	Reduced solar plant capacity		Renewables production	Electricity
Change in climate geographical pattern	Reduced generation capacity of existing renewables	Local increase	Renewables production	Electricity
Side effects from other countries	Biomass, oil and gas import disruptions		Total primary production	Biomass, oil and gas

The situation with renewables is much patchier. The previously mentioned US draft Climate Assessment contains a table of *Challenges in energy supply* (page 181) [5] and in the column "Solar PV Wind", any quantification of trends is dropped by the following comment: "Impacts projected but not well defined at this time". This sort of difficulty comes as no surprise because estimates of renewables productivity strongly depend on local conditions and therefore detailed knowledge of geographical patterns of future climate are needed before any assessment [4]. Concerns about the productivity of renewables are legitimate but there are two big pros to be taken into account: they are almost all immune from water scarcity and solar power, in particular, is available when it is needed most - during peak demand hours.

Last but not least, vulnerability is fed also by the uncertainty of the supply of imported fuel. Changes in climate could spare some countries and heavily hit others in terms of resource endowment. In the latter case, if the affected country is an exporter, the importer could also experience a shortage of fuel.

Beyond mining and production operations, the effects of peculiar weather conditions could jeopardise the energy supply sector in other ways like the transmission infrastructure. Moreover, electricity outages could have widespread effects as electric powered instrumentation, compression pumps and processing equipment are essential links in the process of creating and moving gas to the end customer. In some instances, even the brief, temporary loss of electric power can put a gas production, processing, compression, or storage facility out of service for long periods of time, especially where weather conditions delay access to those facilities [24]. This is difficult to quantify.

Finally, there is an impact on transmission lines. A 5 °C increase in air temperature could decrease transmission line capacity. Estimation in +5 °C scenario suggest a decrease by 7% – 8% [25] which, for a rough and conservative assessment, could imply a 2% loss per °C on the electrical energy.

NET ENERGY BALANCE

After having analysed the effects of climate change both on energy supply and demand, it is now possible to look at the values for each of the elements in the energy balance and find out how rooted is the concern for possible threats to the energy security in Europe. In order to make the effects on energy balance easier to read, a simplified version of the above energy balance is used and all sources together are collapsed in a single column. Supply area is split in three macro areas: primary production, net import and others. Energy demand is split in: industry, transport and residential + services. The total balance is closed with energy transformation, energy management and non energy use.

Breakdown of effects of climate change on energy sector will be the following:

- Energy supply will not be affected by climate change, extreme weather events apart;
- Transformation will suffer a decrease of efficiency in thermal power plant expressed by:

$$\Delta E_{in} = 0.005 \times \Delta T \times E_{in} \tag{2}$$

E_{in} is amount of energy to be transformed and ΔT is the temperature change;

- Electricity distribution will experience an additional loss of energy expressed by:

$$\Delta E_{el} = 0.02 \times \Delta T \times E_{el} \tag{3}$$

where E_{el} is amount of electricity produced and distributed;

- On the demand side, change in consumption will be driven by change in HDD and CDD:

$$E^{heat}_{acc} = E^{heat}_{bcc} \times \_HDD_{acc} / HDD_{bcc} \qquad (4)$$

$$E^{cool}_{acc} = E^{cool}_{bcc} \times \_CDD_{acc} / CDD_{bcc} \qquad (5)$$

In a +2 °C scenario, the previous assessment applied to the EU27 2010 energy balance produces the effects listed in table 4. The green cells highlight values unaffected by climate change while the red cells indicate affected values. Overall, a 75 Mtoe reduction in energy consumption improves the overall energy balance by 4%. This means that, in a +2 °C scenario expected to be far from any tipping point, the issue of energy security for Europe will not be exacerbated but possibly alleviated.

Table 4. Modification of EU 27 energy balance as effect of climate change in +2 °C scenario

		Supply bcc	Consumption bcc	Supply acc	Consumption acc
Supply	Primary	831,105.00		831,105.00	
	Net imports[1]	951,806.00		951,806.00	
	Others supply[2]	-14,497.00		-14,497.00	
Energy transformation /distribution	Transformation losses[3]		366,454.00		372,969.04
	Other transf. losses[4]		22,887.00		22,887.00
	Distribution losses		27,716.00		37,263.44
	Energy industry[5]		97,331.00		97,331.00
Non-energy consumption			114,792.00		114,792.00
Final energy consumption	Industry		289,621.00		289,621.00
	Transport		365,117.00		365,117.00
	Households & services		497,765.00		405,564.76
Total		1,768,414.00	1,781,683.00	1,768,414.00	1,705,545.24

[1] Imports-exports
[2] Recovered products + Stock change - Bunkers + Direct use
[3] (Tranf. input - output) (Main act. and autoprod. of thermal power station + nuclear power stations)
[4] Briquetting, Coke-oven, Blast-furnace and District heating plants + Gas works + Refineries
[5] Exchanges and transfers. Returns + Consumption of the energy branch

CONCLUSION AND MAIN FINDINGS

Energy demand and supply are going to be modified in Europe as an effect of climate change and the possible threat to energy security needs to be discussed in advance. Data of EU27 energy balance in 2010 have been analysed and their change in a +2 °C scenario has been discussed. There are significant differences among countries but this study analyses European data as a whole. Quantitative assessments have been made at the level of first order calculation, both for the sake of simplicity and because the great incertitude on future scenarios could make a supposed higher resolution meaningless. The main findings are:

- More than 60% (industry + transport) of European energy consumption is today (2010) climate independent;
- The remaining 40% of consumption is about households and services. Around 70% of this share goes in heating while less than 5% goes in cooling. Heating and cooling are usually correlated to HDD and CDD which are expected to change in -20% HDD and +65% CDD. The net result will be a reduction of 5% in energy consumption;
- On the supply side, an exact picture is much more difficult to depict but effects are expected to be small if, overall, not zero. Mining and extraction are expected to be climate independent. The only area which could be affected is electricity production and distribution. In fact higher temperature means lower efficiency of thermal power plants and increase in distribution losses. Combination of the two effects results, ceteris paribus, in a decrease of 1-2% in energy availability but the energy sector could easily cope with this request of resilience.

All previous assessments are made without taking into account consequences of extreme weather events, which are expected to increase in intensity and frequency but whose effects on energy sector can hardly be estimated. Tipping points are also taken out of any quantitative assessment.

The overall picture is that climate change in Europe is not going to pose a dramatic challenge to energy security. In fact, the large share of fuel import, more than 70%, will remain the biggest threat for decades to come.

Previous conclusions do not want to underestimate the importance of proactive actions to make the energy sector better prepared to adapt to climate changes and top priority should be given to improving the power sector's resilience. Back-up power generation, additional peak power capacity, distributed generation, interconnections among electric grids and portable generators to critical facilities for possible outages are examples of needed actions. Integration among parts is needed in order to maximise efficiency and flexibility but extreme weather suggests that making each part able to survive any possible disruption is also essential. The World Energy Council (WEC), recently compiled a study along with Cambridge University and the European Climate Foundation, urging generators to examine their vulnerability to climate change, saying that with suitable adaptations the worst of the problems could be avoided. That said, given the usual large incertitude on energy forecast aggravated by the intrinsic limits of climate models, a part of 'play it by the ear' will be unavoidable.

REFERENCES

1. Edwards, P. N., A Vast Machine, Boston: The MIT Press, 2010.
2. Center for Climate and Change Environmental Forecasting, Climate Tipping Points: Current Perspectives and State of Knowledge, Washington: U.S. Department of Transportation, 2009.
3. Ebinger, J. and Vergara, W., Climate Impact on Energy Systems, Washington: The World Bank, pp 224, 978-0-8213-8697-2, 2011.

4. ESPON, Discussion Paper: Impacts of Climate Change on Regional Energy Systems, ESPON, [Online], 29 October 2010, http://www.espon.eu/export/sites/default/Documents/Projects/AppliedResearch/ReR ISK/RERISK-Discussion-Paper-Climate-Change.pdf, [Accessed: 12-November-2013]
5. Dell, J. and Tierney, S., Energy Supply and Use (Draft for Public Comment), NCADAC (National Climate Assessment Development Advisory Committee), 2013.

6. Vautard, R., et al., The European climate under a 2 °C global warming, Environmental *Research Letters,* pp 11, 2014.

7. Eurostat, Energy Trends, [Online], 2014, http://epp.eurostat.ec.europa.eu/statistics_explained/index.php/Energy_trends, [Accessed: 04-April-2014]

8. Wilbanks, T. J., et al., Industry, Settlement and Society, Climate Change 2007: Impacts, Adaptation and Vulnerability, Contribution of Working Group II to the Fourth Assessment Report of the Intergovernmental Panel on Climate Change, Cambridge, UK: M. L. Parry, O. F. Canziani, J. P. Palutikof, P. J. van der Linden and C. E. Hanson (Eds), Cambridge University Press, pp 357-390, 2007.

9. Ball, R., Breed, W. S. and Hillsman, E., Industry, Energy, and Transportation:Impacts and Adaptation, Washington: IPCC, 2011.

10. European Environment Agency, Climate change, Impacts and Vulnerability in Europe 2012, Copenhagen: EEA, EEA Report No 12/2012, 978-92-9213-346-7, 2012.

11. Aebischer, B., Catenazzi, G. and Jakob, M., Impact of Climate change on Thermal Comfort, Heating and Cooling Energy demand in Europe, Centre for Energy Policy and Economics (CEPE), Zurich: ETH Zurich, ECEEE 2007 SUMMER STUDY, 2007.

12. Dowling, P., The impact of Climate Change on the European Energy System, *Energy Policy*, Vol. 60, pp 406-417, 2013.

13. Labriet, M., et al., Worldwide impacts of Climate change on Energy for Heating and Cooling, Mitigation and Adaptation Strategies for Global Change, November 2013.

14. Connolly, D., et al., Heat Roadmap Europe 2050: Second Pre-study for the EU27, Aalborg, Denmark: Department of Development and Planning, Aalborg University, 2013.

15. Economidou, M., et al., Europe's Buildings under the Microscope, Brussels: Buildings Performance Institute Europe (BPIE), ISBN: 9789491143014., 2011.

16. Bertoldi, P., Hirl, B. and Labanca, N., Energy Efficiency Status Report, Luxembourg: Publications Office of the European Union, JRC Scientific and policy report, 2012.

17. Field, C. B., et al., *Summary for Policymakersin: Managing the Risks of Extreme Events and Disasters to Advance Climate Change Adaptation*, Cambridge, UK: Cambridge University Press, Cambridge, United Kingdom and New York, NY, USA., 2012.

18. Polygon Group, Offshore Oil, [Online], 31 January 2013, http://www.polygongroup.com/demo/files/2013/01/31-Polygon_Offshore-Oil-eng.pdf, [Accessed: 12-November-2013]

19. Van Vliet, MTH, et al., Global River Temperatures and Sensitivity to Atmospheric Warming and Changes in River Flow, *Water Resources Research*, Vol. 47, No. 2, 2011.

20. Daycock, C., Jardins, R. and Fennel, S., Generation Cost Forecasting Using On-Line Thermodynamic Models, *Proceedings of the Electric Power*, Baltimore, Md., USA, 2004.

21. Linnerud, K., Mideksa, T. K. and Eskeland, G. S., The Impact of Climate Change on Nuclear Power Supply, *The Energy Journal*, Vol. 32, No. 1, pp 149, 2011.

22. Durmayaz, A. and Sogut, O.S., Influence of Cooling Water Temperature on the efficiency of a Pressurized-water Reactor Nuclear-power plant, *International Journal of Energy Research*, Vol. 30, pp 799-810, 2006.

23. NERC, North America Electric Reliability Corporation, Electricity and Natural Gas Interdependency, 2012.

24. Sathaye, J. A., et al., Estimating impacts of warming temperatures on California's electricity System, *Global Environmental Change*, Vol. 23, No. 2, pp. 499-511,2013.

25. DOW, U.S. Energy sector vulnerabilties to climate change and extreme weather, Washington: US DOE, July 2013.

12

Trends in Residential Energy Consumption in Saudi Arabia with Particular Reference to the Eastern Province

Farajallah Alrashed[*1], *Muhammad Asif*[2]

[1]School of Engineering and Built Environment
Glasgow Caledonian University, Glasgow, UK
e-mail: farajallah.alrashed@gcu.ac.uk
[2]Department of Architectural Engineering
King Fahd University of Petroleum & Minerals, Dhahran, Saudi Arabia

ABSTRACT

Residential buildings are vital in the energy scenario of Saudi Arabia as they account for 52% of the total electricity consumption. The Eastern Province, due to its harsh weather conditions, is one of the most challenging areas in Saudi Arabia in terms of residential energy consumption. The province is vital also because of its large land area, accounting for almost one third of the entire country. This article investigates some of the important factors related to the residential energy consumption i.e. weather conditions, types of dwellings, building envelops, air-conditioning (A/C) systems and domestic appliances especially cooking ovens. The work is based upon an analysis of the actual monthly electricity consumption for 115 dwellings in Dhahran for the year 2012. The investigated buildings include 62 apartments, 28 villas, and 25 traditional houses. The annual average electricity consumption for the surveyed dwellings was found to be 176.5 kWh/m², a value higher than international energy-efficiency benchmarks. It is found that the use of mini-split A/C systems, thermal insulation and double-glazed windows can help reduce the electricity consumption by over 30%.

KEYWORDS

Domestic energy consumption, Air-conditioning systems, Thermal insulation, Window glazing systems, Cooking energy, Saudi Arabia, Dhahran.

INTRODUCTION

The building industry has a key role to play in achieving sustainable development in any country [1]. Buildings contribute to environmental issues ranging from the excessive use of resources during the construction and the operation stages to polluting the surrounding environment [2]. Buildings not only use resources such as energy and raw materials but they also generate waste and potentially harmful atmospheric emissions [3]. Buildings are responsible for a substantial proportion of the global greenhouse gases (GHGs) emissions. For example, according to the United Kingdom Government, around 40% of the national energy consumption and carbon dioxide emissions (CO_2) are associated with buildings [4]. Buildings account for 40% of the total energy consumption at the European Union (EU) level as well [5]. In the wake of such a crucial role of buildings in its energy and environmental scenario, the EU through its 20-20-20 Directive has set 20-20-20 targets. Specifically, these targets aim to achieve by 2020 a 20% reduction in EU greenhouse gas (GHG) emissions from 1990 levels; raising the

---

[*] Corresponding author

share of EU energy consumption produced from renewable resources to 20% and a 20% improvement in the EU's energy efficiency [6]. Similarly, the United States (US) Green Building Council [7] suggests that the US commercial and residential building sector accounts for 39% of CO_2 emissions per year, more than any other sector in the country. According to Alnaser *et al.* [8], construction and operation of buildings have an enormous direct and indirect impact on the environment. The annual environmental impact of the global building sector includes energy use (42%), atmospheric emissions (40%), raw materials use (30%), solid waste (25%), water use (25%), water effluents (20%), land use (12%), and other emissions (13%) [8]. Given the massive growth in new construction and the inefficiencies of existing building stock worldwide, in a business as usual scenario, the level of GHGs emissions from buildings is set to rise in future [3]. If the desired targets for GHGs emissions reduction are to be met, emissions from the building sector need to be tackled with much greater seriousness and vigour than the past efforts and in this respect energy-efficient and sustainable buildings are critical to be promoted. In order to develop robust strategies to stimulate the take up of energy-efficient buildings, it is crucial to have a thorough understanding of current practices and future trends in the building sector.

ENERGY CONSUMPTION IN SAUDI ARABIA

The total installed electricity generation capacity in Saudi Arabia is 44,485 MW, all being supported by oil and natural gas with a respectively share of 57% and 43% [9]. In the wake of fluctuating oil prices in recent years, natural gas has seen a jump in its share in electricity production - the contribution from natural gas has increased from 37% in 2007 to 43% in 2009 [10].

The demand for electricity is experiencing a rapid growth in Saudi Arabia. Since 1990, for example, the demand has increased at an annual rate of 6% [9]. Due to the economic and population growth, statistics suggest that, in comparison to 2008, the primary energy demand is expected to increase by 50% in 2020 [11]. Furthermore, the per capita electricity consumption is also increasing rapidly due to factors like urbanization, subsidized tariffs and increased use of energy intensive appliances as shown in Figure 1a [9]. The residential sector is the biggest consumer of electricity - presently it accounts for 52% of the total national electricity consumption [9] as indicated in Figure 1b and it is expected that by year 2025 the demand from this sector would double [12]. The rapid growth in electricity demand is largely due to inefficient use of electricity which in turn is associated with extremely subsidized tariffs as also highlighted by Alyousef and Stevens [13]. To respond to the electricity growth trend, the country needs to take appropriate initiatives not only to boost its power generation capacity but also to make residential sector more energy efficient.

An analysis of the construction sector suggests that most of the projects being undertaken are residential buildings in order to meet the demand for new homes - the statistics provided by the Ministry of Municipal and Rural Affairs indicate that the majority of licenses issued for construction in Saudi Arabia are for residential buildings [14]. In addition, the residential sector is set to experience a similar growth in future as the Saudi population is rising at a rate of 2.5% per year and only 24% of the Saudi nationals have their own homes [15]. Estimates suggest that around two-third of the population is under the age of 30 years [16]. To meet the needs of the constantly growing population, the country needs to build 2.32 million new homes by 2020 [17].

In a survey undertaken by the Government, it was discovered that about 60% of the total electricity consumed in summer goes into air conditioning (A/C) systems [18]. According to the Saudi Ministry of Water and Electricity [9], the electricity consumption in the country has increased by 35% over the last two decades largely due to intensive use of air conditioning in summer. It is therefore crucial for Saudi Arabia to improve the energy consumption trends in residential buildings and to move towards energy efficient buildings. This paper aims to discuss the energy trends in the Saudi residential sector based upon a survey of the actual monthly electricity consumption for 115 housing units in the Eastern Province.

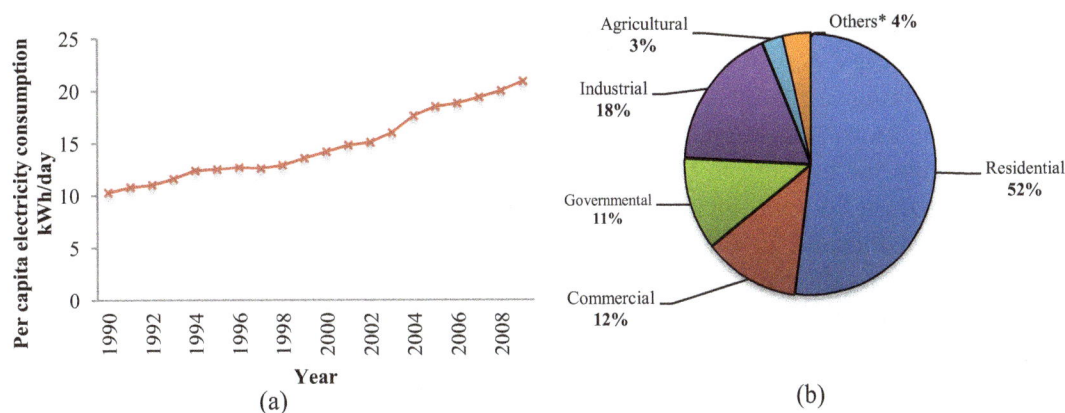

*Hospitals, mosques, streets, and charity associations

Figure 1. Per capita electricity consumption in Saudi Arabia (a); Electricity consumption by sector (b)

Saudi Arabia can be classified climatically into five different inhabited climatic zones: Subtropical with a Mediterranean subzone and Mountainous subtype (for example Khamis Mushait); Hot-Dry with a Maritime Desert subzone (for example Jeddah); Hot-Dry Maritime subzone (for example Dhahran); Cold-Dry with a Desert subzone (for example Quriat), and Hot-Dry with a Desert subzone (for example Riyadh) [19]. The Eastern Province is a vital region in Saudi Arabia because of its large land area, accounting for almost one third of the entire country. Due to its harsh weather conditions, it is one of the most challenging areas in Saudi Arabia in terms of residential energy consumption. Dhahran is a main representative city in the Eastern Province which is subject to Hot-Dry Maritime weather subzone with the maximum temperature highest among all climatic zones (see Table 1).

Table 1. Temperature comparison between Saudi represented climatic zones [20]

Location	Maximum dry-bulb temperature (°C)	Mean dry-bulb temperature (°C)
Dhahran	45.7	25.8
Guriat	43.9	19.8
Riyadh	43.7	25.12
Jeddah	41.7	27.9
Khamis Mushait	34.3	18.9

SURVEY

To investigate the energy consumption trends in residential buildings, the study undertook a survey of dwellings in the Dhahran region on the basis of monthly electricity bills. Through the questionnaire based survey, randomly selected dwellings were investigated for their monthly electricity consumption from January to December 2012.

The survey also looked into eight main electricity consumption related features: dwelling type, dwelling age, dwelling conditions area, type of A/C system, thermal insulation, window glazing system, and fuel for cooking. A total of 128 responses were received of which 13 were rejected for incomplete information. The survey data was gathered through in-person approach using various modes including interviews of residents/inhabitants, investigation of building plans and electricity bills and walk through the buildings.

In hot-humid climate, similar to Dhahran, A/C and building envelop significantly influence the energy consumption in buildings [21-23]. In Saudi dwellings, the A/C system and cooking are the two most important energy consuming factors, collectively accounting for more than 80% of total household energy consumption [18-24]. The envelope system for housing in Dhahran region is typically made of concrete while the external walls from hollow-block and the roof are made of rapid-slab/Hordi-slab [25]. In Saudi dwellings, thermal insulation is usually applied in external walls and is unlikely to be applied within roofs and slab on grade [23-26]. The most common thermal insulation materials are Polystyrene and Polyurethane with a typical thickness of 5 centimeters [27]. Due to lack of data on the type of insulation used in the surveyed dwellings, all of them are assumed to have similar insulation while it is noteworthy that both of the aforementioned insulations are quite similar in performance [27, 28].

RESULTS AND ANALYSIS

The types of dwelling studied in the survey are apartments, traditional houses and villas representing over 90% of the dwellings in the Eastern Province [29]. A traditional house in Saudi Arabia is a dwelling that has at least one external wall is shared with a neighbour and does not have a fence (i.e. 100% of the land is built). The villa is a free-standing (detached) residential building surrounded by fence. Generally, the type of dwelling for more than half of the responses is apartment (see Figure 2a). Over half of the surveyed dwellings were built within the last ten years (see Figure 2b).

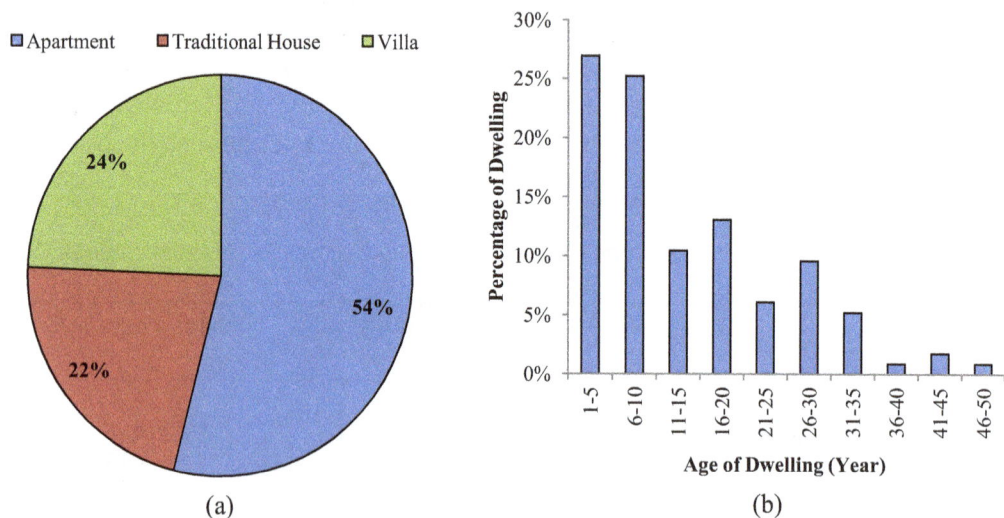

Figure 2. Surveyed dwellings by type (a); Percentage of surveyed dwellings in terms of age (b)

The survey results indicate that the average conditioned area for the surveyed dwellings is about 200 m². Of these, apartments and villas respectively account for the smallest and the largest conditioned areas (see Figure 3a). It is also observed that the conditioned area for the surveyed villas and traditional houses - all built over the last three decades - has increased in recent years as shown in Figure 3b. On the other hand, the conditioned area for apartments has decreased during the last two decades (see Figure 3b).

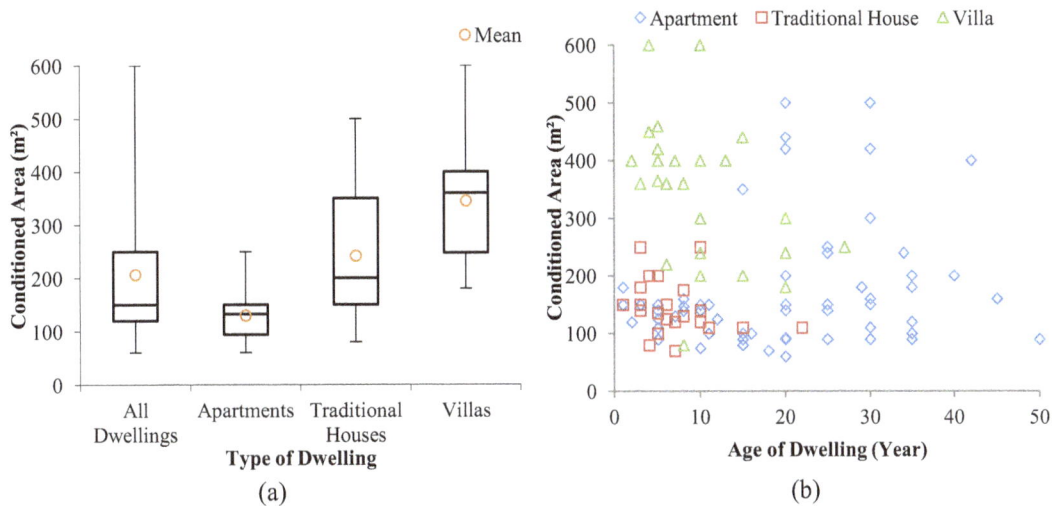

Figure 3. Box-Plot of the conditioned area for all type of dwellings (a); Conditioned area for all dwellings in terms of age (b)

The survey results reveal that about 92% of the surveyed dwellings are using mini-split and window-type (conventional type of A/C units that are fixed in a window frame through the walls) A/C systems (see Figure 4a). It is also observed that compared to mini-split and window-type systems, use of central systems is a recent phenomenon (see Figure 4b). It is also observed that apartments and villas employ window-type and mini-split systems, while the traditional houses mainly go for the window-type systems. It is also seen that the central systems are mainly used in apartments and villas as none of the surveyed traditional houses employed this type of system.

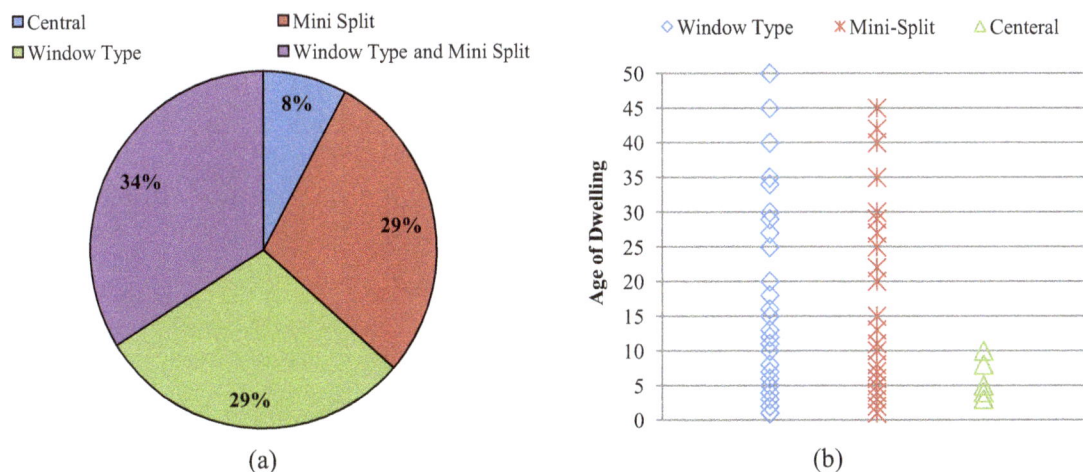

Figure 4. Surveyed dwellings by type of A/C systems (a); Surveyed A/C systems in terms of age (b)

The survey results indicate that the thermal insulation is used within 61% of the surveyed dwellings. It is also observed that the use of thermal insulation started two decades ago and none of the dwellings built over the last five years is thermally un-insulated (see Figure 5a). It was revealed that more than half of the apartments and most of the villas were thermally insulated. On the other hand, more than 75% of the traditional houses were without insulation (see Figure 5b). Moreover, it was found that dwellings employing window-type A/C systems are mostly without thermal insulation.

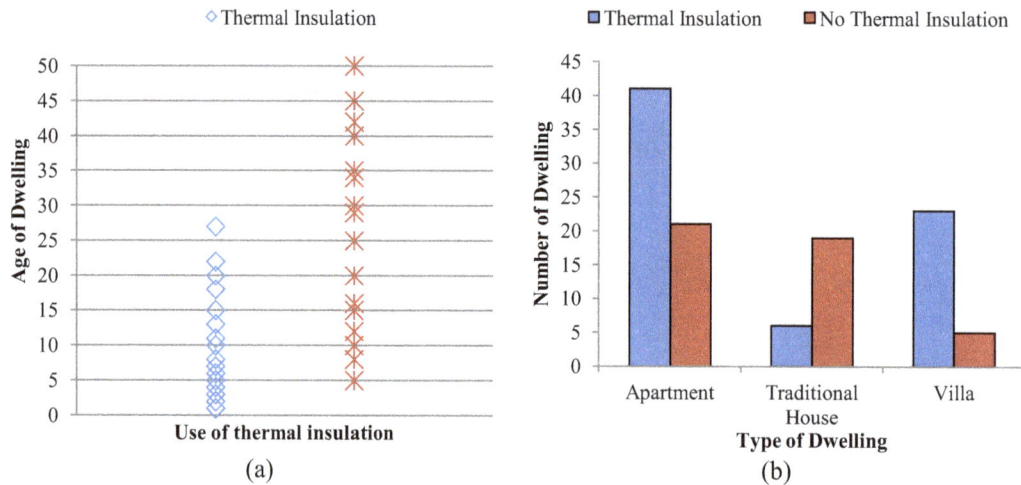

Figure 5. The use of thermal insulation in terms of age (a); The use of thermal insulation in terms of type of dwelling (b)

In terms of the window glazing system, the main focus was on the number of glazing layers. It was found that double-glazed system is used within 50% of the surveyed dwellings. The survey results show that while the use of double-glazed system has increased over the last decade, single-glazed windows are still being applied (see Figure 6a). It was also observed that more than two-thirds of the villas and more than half of the apartments used double-glazed windows, while only 4 of the 25 traditional houses surveyed in the study were using double-glazed system (see Figure 6b).

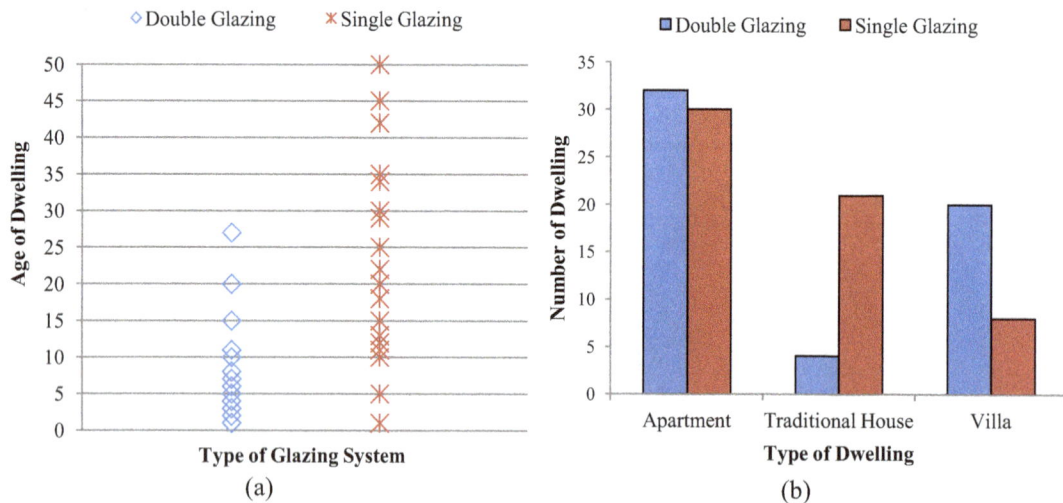

Figure 6. Surveyed type of glazing systems in terms of age (a); Surveyed type of glazing systems in terms of type of dwelling (b)

In terms of cooking fuels, both gas and electricity are used in Saudi dwellings. About 60% of the responses in this study suggested the use of gas for cooking while the rest used electricity. It is also found that the use of electricity for cooking has become common over the last decade (see Figure 7a). The survey results also revealed that electric cooking is more common in apartments as about 60% of the apartment responses indicated to be using it while most of the villas and traditional houses suggested to be using gas (see Figure 7b).

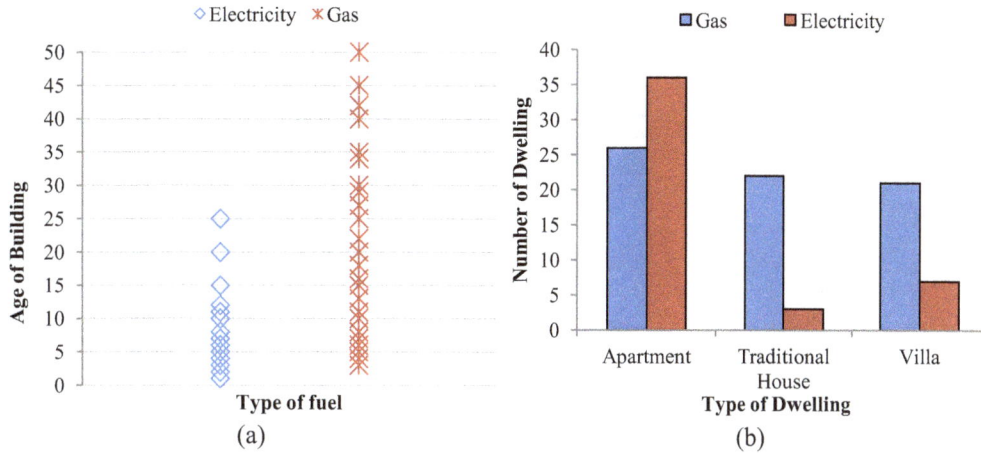

(a)   (b)

Figure 7. Surveyed type of fuel for cooking in terms of age (a); Surveyed type of fuel for cooking in terms of type of dwellings (b)

The overall electricity consumption for the surveyed dwellings is found to be between 27 and 401 kWh/m²/year whereas the average value is calculated to be 176.5 kWh/m²/year. The electricity consumption for 43% of all dwellings is reported to be between 125 and 174 kWh/m²/year (see Figure 8a). In terms of type of dwellings, the average electricity consumption for apartments, traditional house, and villas is respectively, 196.5, 156.5 and 150 kWh/m²/year (see Figure 8b). It is observed that the annual electricity consumption per square meter for apartments tends to be higher than the other types.

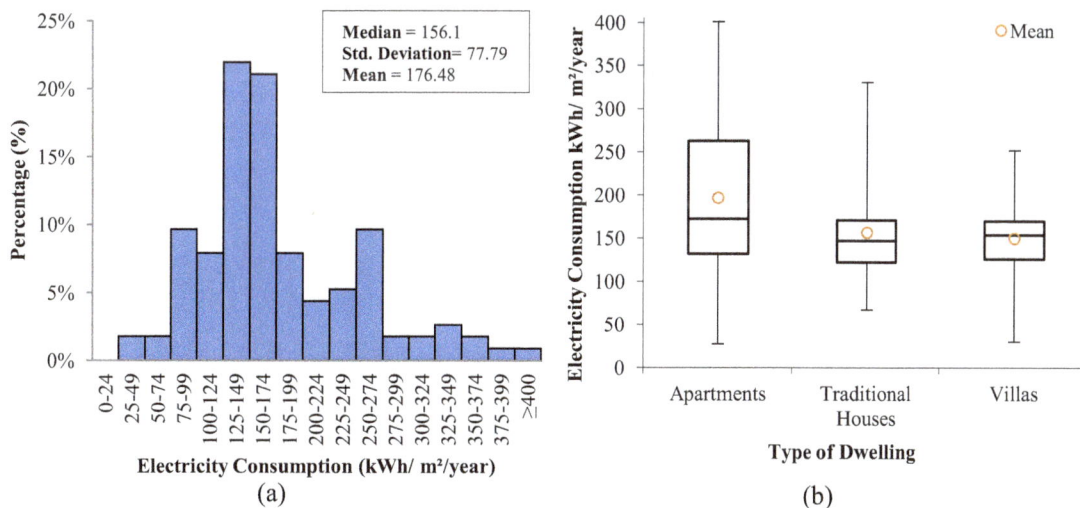

(a)   (b)

Figure 8. Distribution of electricity consumption for all survey dwellings (a); Box-Plot of electricity consumption for all type of dwellings (b)

Furthermore, it was observed from the analysis that the average of electricity consumption for dwellings using electricity as source of energy for cooking is about 34% much higher than dwelling that uses gas (see Figure 9a). The average of electricity consumption for dwellings with mini-split system was found to be almost half the electricity consumption for dwellings with central system or window-type system (see Figure 9b).

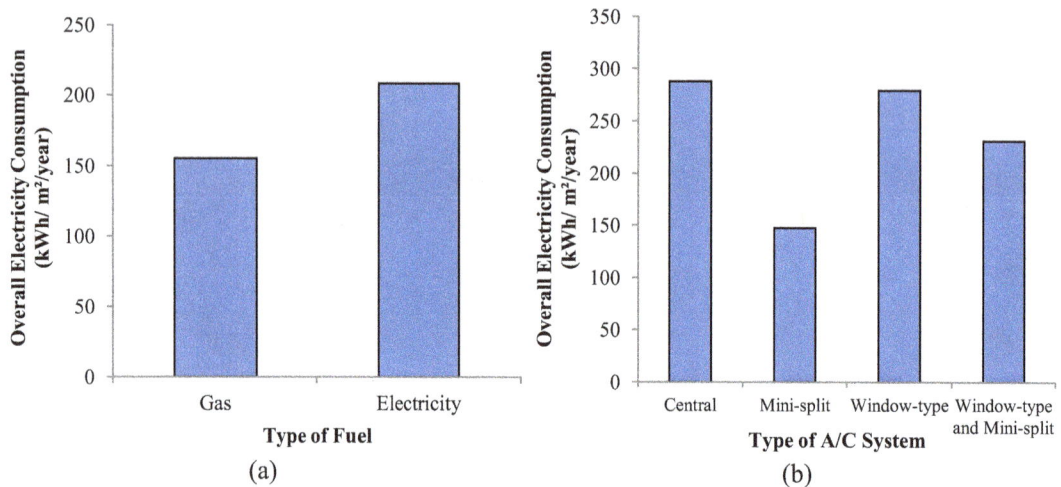

Figure 9. Average electricity consumption in terms of cooking fuel (a); Average electricity consumption in terms of type of A/C system (b)

The average electricity consumption for dwellings with thermal insulation is about 32% lower than the dwellings without thermal insulation (see Figure 10a). About 87% of the surveyed dwellings with thermal insulation employed double-glazed window systems. The average electricity consumption for dwellings with double-glazed windows is about 35% lower than the dwellings with single-glazed window (see Figure 10b). Around 94% of dwellings with double-glazed windows are thermally insulated.

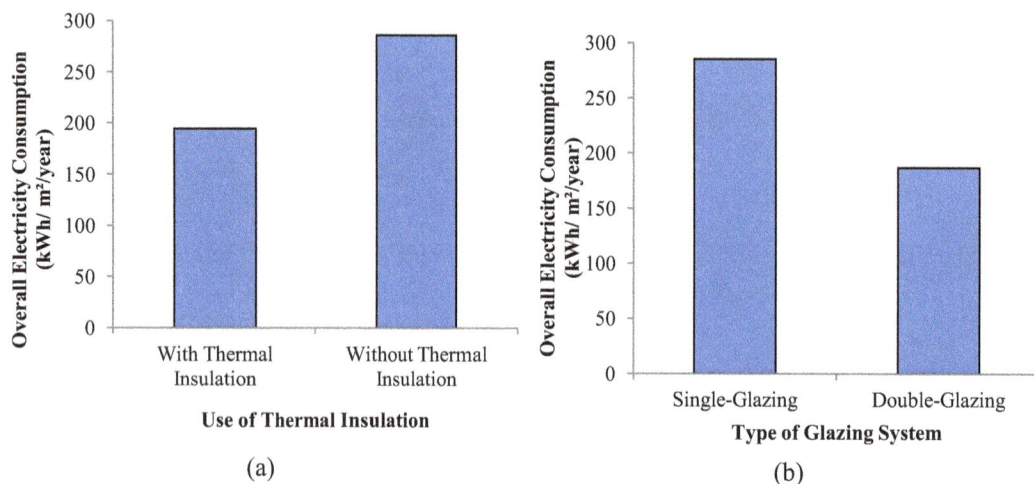

Figure 10. Average electricity consumption in terms of use of thermal insulation (a); Average electricity consumption in terms of type of glazing system (b)

In terms of overall electricity consumption, the survey results reveal that the annual electricity consumption per square meter especially for apartments and traditional houses decreases when the conditioned area increasing and vice versa (see Figure 11a). It is also observed that the electricity consumption of new homes tend to be higher in comparison to the old ones (see Figure 11b).

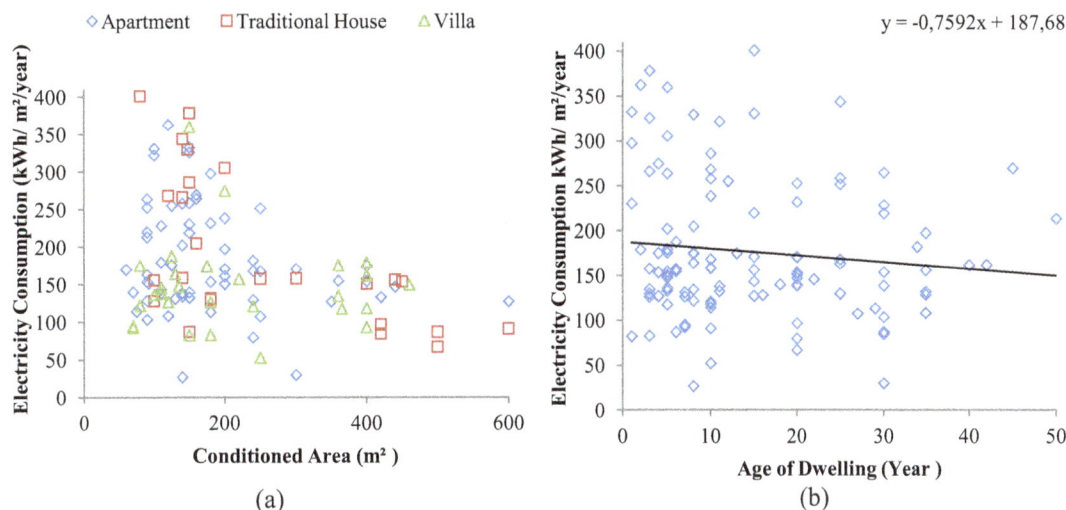

Figure 11. Annual electricity consumption per square meter for all type of survey dwellings (a);
Linear regression for electricity consumption with age of dwelling (b)

DISCUSSION AND CONCLUSIONS

The study highlights some energy consumption and efficiency trends in the residential sector in Dhahran region of Saudi Arabia. Among the three dwelling types in this study (i.e. apartments, traditional house and villa), villas and traditional houses tend to have lower electricity consumption per square meter compared to apartments. However, the apartment average conditioned found out to be 130 m² while villa and traditional house found to be 345 and 240 m² respectively. The average of electricity consumption for villas and traditional houses are close, despite most of the surveyed traditional houses were using window-type system, single-glazed windows, and un-insulated external walls. This could be due to the fact that traditional houses have less exposed area to the harsh weather conditions and more shading on external walls.

In Saudi Arabia, four types of A/C systems are mainly used in residential buildings: window-type, mini-split, central and evaporative cooler [25]. Given the fact that the Eastern Province is subject to high humidity [20], the use of evaporative cooling system is limited. While the A/C in the Eastern Province is largely dominated by window-type and mini-split systems, the use of central systems has started to emerge over the last few years. The survey results reveal that dwellings with mini-split system have about half the annual electricity consumption for dwellings using window-type and central systems. However, only 29% of the surveyed dwellings are using mini-split system, and 34% of them use mini-split in conjunction with window-type systems.

The application of thermal insulation and double-glazing systems has become common over the last 20 years. This might be due to the encouragement by the Saudi Government through increasing the public awareness about the importance of energy efficiency and thermal insulation. In 1985, the Saudi Government started to pay attention to the significance of thermal insulation and its impact on the energy saving [11]. Subsequently a number of policy initiatives have been taken to promote the cause. For example, in 1994, it was made mandatory for all government projects to have thermal insulation [30]. Thereafter, in 2010, it became a mandatory requirement for all new dwelling to have insulated external walls [31]. These initiatives appear to have a positive impact on the increased use of thermal insulation across the board including the residential sector.

It is also observed that the use of electric cooking has become common over the last decade. Although, the use of natural gas is less harmful for the environment in comparison with fossil fuel based electricity, the use of the latter is becoming more and more common for cooking needs [32]. Presently, there is no gas pipeline network in Saudi Arabia and the provision of gas is through refillable cooking gas cylinders, and this could be an important factor in the promotion of electric cooking in recent years.

The annual average electricity consumption for the dwellings surveyed in this study was found to be 176.5 kWh/m². In a global perspective, this value appears to be quite high. For example, it is higher than its corresponding value in the US's hot-humid climate (i.e. the southern part of the US), which is about 130 kWh/m² [33]. It is also higher than the international benchmark values for energy-efficient residential buildings - the maximum annual primary energy demand for passive houses being120 kWh/m² [34] and about 90 kWh/m² for Zero-Energy Homes [35].

In order to increase the energy-efficiency in the Saudi residential sector some energy issues are required to be tackled further. It has been indicated from this survey that the use of mini-split A/C system, thermal insulation in external walls, and double-glazed windows can help in reducing the electricity consumption by over 30%. The use of mini-split A/C systems in dwellings should be encouraged and other types of A/C especially the central system should be avoided. The stimulation program to use thermal insulation should stay alive and attention should be paid to other factors such as the application of thermal insulation in roof and the use of double-glazing windows.

REFERENCES

1. GhaffarianHoseini, A., Dahlan, N. D., Berardi, U., GhaffarianHoseini, A., Makaremi, N. and GhaffarianHoseini, M., Sustainable Energy performances of Green Buildings: A review of Current Theories, Implementations and Challenges, *Renewable and Sustainable Energy Reviews,* Vol. 25, pp 1-17, 2013.

2. Hussin, J., Abdul Rahman, I. and Memon, A., The way Forward in Sustainable Construction: Issues and Challenges, *International Journal of Advances in Applied Sciences,* Vol. 2, No. 1, pp 31-42, 2013.

3. UNEP SBCI, *Buildings and Climate change: A Summary for Decision-makers*, Paris: United Nation Environmental Programme, Sustainable Buildings and Climate Initiative, 2009.

4. Department for Communities and Local Government, *Improving the Energy efficiency of our Buildings: A guide to Energy Performance Certificates for the Marketing*, Sale and let of Dwellings, London, 2014.

5. Europa, Summaries of European Union Legislation: Energy Performance of Buildings, http://europa.eu/legislation_summaries/internal_market/single_market_for_goods/c onstruction/en0021_en.htm, [Accessed: 26-May-2014].

6. Europa, Summaries of European Union Legislation: Promotion of the use of energy from renewable sources, http://europa.eu/legislation_summaries/energy/renewable_energy/en0009_en.htm, [Accessed: 26-May-2014]

7. US Green Building Council, Building and Climate change, http://www.documents.dgs.ca.gov/dgs/pio/facts/LA workshop/climate.pdf. [Accessed: 19-Aug-2013]

8. Alnaser, N. W., Flanagan, R. and Alnaser, W. E., Model for calculating the Sustainable Building Index (SBI) in the Kingdom of Bahrain, *Energy Build.*, Vol. 40, No. 11, pp 2037-2043, 2008.

9. Ministry of Water and Electricity, *Electricity: Growth and Development in the Kingdom of Saudi Arabia*, Riyadh, 2009.

10. Reegle, Country Energy Profile: Saudi Arabia - Clean Energy Information Portal, http://www.reegle.info/countries/saudi-arabia-energy-profile/SA,   [Accessed: 08-Jul-2013]

11. Alyousef, Y. and Abu-eid, M., Energy efficiency Initiatives for Saudi Arabia on Supply and Demand Sides, *Energy Efficiency - A Bridge to Low Carbon Economy*, 1st ed., Z. Morvaj, Ed. InTech, 2012, pp 279-308.

12. Obaid, R. R. and Mufti, A. H., Present State, Challenges, and Future of Power Generation in Saudi Arabia, *IEEE Energy 2030 Conference*, 2008, pp 1-6.

13. Alyousef, Y. and Stevens, P., The cost of Domestic Energy prices to Saudi Arabia, *Energy   Policy*,   Vol.   39,   No.   11,   pp   6900-6905,   2011.

14. Ministry   of   Municipal   and   Rural   Affairs,   Statistics, http://www.momra.gov.sa/GeneralServ/statistics.aspx, [Accessed: 27-Jun-2013].

15. Deloitte, *GCC Powers of Construction 2010: Construction sector overview*, New York, 2010.

16. Central Department of Statistics & Information, Population statistics database, http://www.cdsi.gov.sa/english/index.php?option=com_docman&task=cat_view&gid=43&Itemid=113, [Accessed: 02-Mar-2013]

17. Sidawi, B., Hindrances to the Financing of affordable Housing in Kingdom of Saudi Arabia, *Emirates Journal for Engineering Research*, Vol. 14, No. 1, pp 73-82, 2009.

18. Khair-El-Din, A.-H. M., Energy Conservation and its Implication for Architectural Design and Town planning in the Hot-arid Areas of Saudi Arabia and the Gulf States, *Solar   and   Wind   Technolgy*,   Vol.   7,   No.   2-3,   pp   131-138,   1990.

19. Said, S. A., Habib, M. and Iqbal, M., Database for building Energy Prediction in Saudi Arabia, *Energy Conversion and Managment*, Vol. 44, No. 1, pp 191-201, 2003.

20. Meteonorm,   Meteonorm: Station   Map, http://meteonorm.com/products/meteonorm/stations/, [Accessed: 08-Aug-2013]

21. Mohammed M. A. and Budaiwi, I. M., Strategies for Reducing Energy Consumption in a Student Cafeteria in a Hot-humid Climate: A Case Study, *Journal os Sustainable Development of Energy, Water and Environment Systems*, Vol. 1, No. 1, pp 14-26, 2013.

22. Al-Mofeez, I. A., Electrical Energy Consumption pre and post Energy Conservation Measures: A Case Study of One-story House in Dhahran, Saudi Arabia, *Journal of the King Saud University*, Vol. 19, No. 2, pp 1-12, 2007.

23. Ahmad, A., Energy Simulation for a Typical House built with Different types of Masonry Building Materials, *Arabian Journal for Science and Engineering*, Vol. 29, No. 2B, pp 113-126, 2004.

24. Dincer, I., Hussain, M. M. and Al-Zaharnah, I., Energy and Exergy use in Residential Sector of Saudi Arabia, *Energy Sources*, Vol. 26, No. 13, pp 1239-1252, 2004

25. Shash, A. and Al-Mulla, E., Major Components of 'Typical Villa' in Saudi Arabia for Price/cost Index Development, *The 6th Saudi Engineering Conference*, Vol. 1, pp 47-62, 2002.

26. Al-saadi, S. N. and Budaiwi, I. M., Performance-based Envelope Design for Residential Buildings in Hot Climates, *Building Simulation 2007*, pp 1726-1733, 2007.

27. Ahmad, E. H., Cost Analysis and Thickness Optimization of Thermal Insulation Materials used in Residential Buildings in Saudi Arabia, *The 6th Saudi Engineering Conference*, Vol. 1, pp 21-32, 2002.

28. Alajlan, S. A., Smiai, M. S. and Elani, U.A., Effective Tools toward Electrical Energy Conservation in Saudi Arabia, *Energy Conversion and Managment,* Vol. 39, No. 13, pp 1337-1349, 1998.

29. Central Department of Statistics & Information, Housing Census Results, http://www.cdsi.gov.sa/2010-07-31-07-00-05/cat_view/31-/138----/342---1431-201 0/341---, [Accessed: 18-Jun-2013]

30. Ministry of Water and Electricity, The Application of Thermal Insulation is Mandatory   for   all   new   Gernmental   Buildings, http://www.mowe.gov.sa/files/forms/Tazlel/Tazlel1.pdf

31. Ministry of Water and Electricity, The Application of Thermal Insulation is Mandatory   for   all   new   Buildings, http://www.mowe.gov.sa/files/forms/Tazlel/Tazlel2.pdf

32. Energy Information Administration, Natural Gas 1998: Issues and Trends, http://www.eia.gov/pub/oil_gas/natural_gas/analysis_publications/natural_gas_1998 _issues_trends/pdf/it98.pdf, [Accessed: 11-Sep-2013]

33. Energy Information Administration, Residential Energy Consumption Survey (RECS),   http://www.eia.gov/consumption/residential/index.cfm,   [Accessed: 19-Aug-2013]

34. Passive   House   Institute,   Passive   House   requirements, http://www.passiv.de/en/02_informations/02_passive-house-requirements/02_passiv e-house-requirements.htm, [Accessed: 26-Jun-2013]

35. Tucson/Pima County, Tucson / Pima County Net-Zero Energy Building Standard: Cash Flow positive on Day 1, 2012.

13

Digestate from Biogas Plants is an Attractive Alternative to Mineral Fertilisation of Kohlrabi

Tomas Losak[*1], *Jaroslav Hlusek*[1], *Andrea Zatloukalova*[2], *Ludmila Musilova*[2], *Monika Vitezova*[2], *Petr Skarpa*[2], *Tereza Zlamalova*[2], *Jiri Fryc*[3], *Tomas Vitez*[3], *Jan Marecek*[3], *Anna Martensson*[4]

[1]Department of Environmentalistics and Natural Resources, Faculty of Regional Development and International Studies, Mendel University in Brno, Brno, Czech Republic
e-mail: losak@mendelu.cz
[2]Department of Agrochemistry, Soil Science, Microbiology and Plant Nutrition, Faculty of Agronomy, Mendel University in Brno, Brno, Czech Republic
[3]Department of Agricultural, Food and Environmental Engineering, Faculty of Agronomy, Mendel University in Brno, Brno, Czech Republic
[4]Department of Soil and Environment, Swedish University of Agricultural Sciences, Uppsala, Sweden

ABSTRACT

This study examined the potential for the use of digestate from biogas plants for the fertilisation of kohlrabi. Kohlrabi was grown in two pot experiments in consecutive years using digestate, mineral fertiliser (urea) with a nitrogen (N) content equivalent to that in the digestate, mineral fertiliser with N, phosphate (P), potassium (K) and magnesium (Mg) contents equivalent to the digestate, and an unfertilised control. At harvest, the soil receiving the digestate application had higher P, K and Mg contents than the control and the urea treatment. The soil N_{min} content was balanced in all fertilised treatments. Soil pH was unaffected by all treatments. Kohlrabi bulbs from the unfertilised control had the lowest weight, nitrate content and ascorbic acid content. Digestate and NPKMg fertiliser treatments increased bulb weight compared with the N-only urea treatment. Ascorbic acid content did not differ between fertilised treatments. There were no differences in bulb nitrate content between the mineral fertiliser treatments, but digestate application gave a low nitrate content. Bulb macroelement contents varied irregularly among treatments.

KEYWORDS

Biogas plant, Digestate, Fertilisation, Soil, Kohlrabi, Yields, Quality.

INTRODUCTION

In recent years there has been a major expansion in the production and use of biogas in Europe for the co-generation of electricity and heat. For farmers, biogas stations (BGS) offer a new and predictable production of 'environmentally friendly' energy. However, the widespread production of biogas presents a number of new questions, including the subsequent use of the residue from anaerobic fermentation - digestate [1, 2]. Field and pot trials to date report positive effects of digestate application to arable land in terms of yield [3-5] or no significant effects [6, 7]. Experts are divided in their opinions on the properties and possibilities for the practical use of digestate as an organic fertiliser [8-10].

---

[*] Corresponding author

Digestion is associated with large losses of organic carbon (C) [11]. During the digestion process, 24-80% of organic dry matter is transformed to methane (CH_4) and carbon dioxide (CO_2) [12]. However, the digestate produced is rich in N and has a high ammonium-N/total N ratio, making it potentially suitable as a fertiliser. Comparisons between digestate applications and mineral N fertilisers based on equivalent amounts of total N have shown lower fertiliser N-values than the mineral N fertilisers [13]. Higher ammonia losses after spreading digested slurry were observed, because anaerobic digestion of manure increases the ammonium (NH_4^+) concentration as well as the pH, and both these factors promote gaseous N loss [14]. Where anaerobic digestate has been applied in the field a significant negative correlation between net N-mineralisation and mineralised-C has been reported in the soil [15]. The digestion of energy crops alone or in combination with animal manures leads to an additional quantity of organic manure [16].

Although current regulations define digestate as an organic fertiliser, its composition and properties are closer to that of compound mineral fertiliser. When fertilising with digestate it is therefore necessary to apply other sources of primary (labile) organic matter of good quality to the soil at the same time, e.g. by ploughing down all post-harvest residues, applying farm manure, compost, straw [1, 9] or using a catch crop, e.g. clover-grass mixture.

The aim of this study was to evaluate the effect of digestate application on chemical soil properties, yield and chemical composition of kohlrabi.

MATERIAL AND METHODS

Two pot experiments were established on 8 June 2010 and on 21 May 2011. Mitscherlich plant pots were filled with 6 kg of medium heavy soil characterised as fluvial soil with the chemical properties as shown in Table 1.

Table 1. Chemical characteristics of soil prior to establishment of the experiments (Mehlich III)

pH/CaCl$_2$	mg/kg			
	P	K	Ca	Mg
7.5	34	159	6,262	303
Alkaline	Low	Satisfactory	Very high	Good

The experiments involved four treatments, as shown in Table 2. The digestate (C/N ratio 4/1) was obtained from a biogas station which uses pig slurry and maize silage from hybrid KWS 1393 as its raw material inputs.

In addition a bio-enzymatic preparation (APD BIO GAS) was applied into the fermentor to increase biogas production. The preparation consists of a mixture of bacterial and enzymatic cultures and nutrients and is claimed to have a significant beneficial effect on the activity of the microorganisms involved in the process of methanogenesis. This results in an increase in biogas production, the elimination of odour due to a decrease in the volatilisation of ammonia, higher homogeneity of the substrate preventing the creation of a surface crust and sediments, and improved handling of the fermentation residue and its application to the land.

Tables 3 and 4 show the composition of the digestate in terms of content of nutrients and hazardous elements, respectively. The contents of all hazardous elements were below the limit specified in Regulation No. 271/2009 Coll. (Table 4).

Mineral fertilisers and digestate were applied by watering and were thoroughly mixed with the entire contents of the pot. Two seedlings of kohlrabi cv. Moravia were planted

10 days after fertilisation. The pots were watered to 60% of maximum capillary capacity and were kept free of weeds. The bulbs were harvested at full maturity on 2 August 2010 in the first year and on 19 July 2011 in the second. Immediately after harvest the individual bulbs without leaves were weighed. Nitrate concentration (mg NO_3^-/kg) was determined in the fresh matter of bulbs with a potentiometer using an ion selective electrode (ISE). The ascorbic acid content was determined in fresh matter using the capillary isotachophoresis method.

The soil was extracted according to Mehlich III (CH_3COOH, NH_4NO_3, NH_4F, HNO_3 and EDTA). The content of available P in the extract was determined colorimetrically and the content of available K, Mg and Ca using an AAS. The ion-selective electrode (ISE) was used to determine the pH value [17]. The content of $N-NH_4^+$ in the soil was assessed colorimetrically; the $N-NO_3^-$ content using ISE. The content of macrobiogenic elements in the plant biomass was assessed after wet mineralisation ($H_2SO_4+H_2O_2$). The N content was determined by Kjeldahl analysis, while colorimetric analysis was used to assess P, and the AAS method to assess K, Ca and Mg [18].

The results were processed statistically using variance analysis followed by testing according to Scheffe ($P < 0.05$).

Table 2. Treatments used in the pot experiment

Treatment No.	Description	Dose of nutrients (g/pot): N-P-K-Mg	Fertiliser
1	Untreated control	0	-
2	N fertiliser	1.5-0-0-0	Urea
3	Digestate	1.5-0.18-0.69-0.08	Digestate
4	NPKMg fertiliser	1.5-0.18-0.69-0.08	Urea, triple superphosphate, KCl, $MgSO_4$

Table 3. Contents of major plant nutrients in the digestate

%	Nutrient				
	N	P	K	Ca	Mg
Of dry matter	11.4	1.37	5.2	2.02	0.62
Of fresh matter	0.72	0.09	0.31	0.13	0.04

Table 4. Contents of hazardous elements in the digestate and max. permissible levels

	mg/kg dry matter								
	Cd	Pb	Hg	As	Cr	Cu	Mo	Ni	Zn
Content	0.1	2.4	0.2	0.2	9.1	99	4.6	8.6	481
max.*	2	100	1	20	100	250	20	50	1200

*maximum permissible level according to Regulation No. 271/2009 Coll.

RESULTS AND DISCUSSION

Post-harvest changes in soil chemical properties

Plant nutrients are present in inorganic plant-available forms to a markedly higher level in digested residue than in untreated waste [19], due to the mineralisation of large amounts of organically bound nutrients during the digestion process [20]. Table 5 shows the post-harvest values of some chemical characteristics of the soil (pH, available P, K, Ca, Mg) following the different treatments.

Table 5. Post-harvest content of available nutrients and pH of the soil receiving the different treatments (2010 and 2011)

Treat. No.	Scheme	pH/CaCl$_2$	mg/kg			
			P	K	Ca	Mg
			Year 2010			
1	Control	7.49[a]	28[a]	142[a]	6,158[a]	275[a]
2	N fertiliser	7.48[a]	25[a]	138[a]	6,153[a]	271[a]
3	Digestate	7.49 [a]	48[b]	162[b]	6,117[a]	302[b]
4	NPKMg fertiliser	7.47[a]	50[b]	151[b]	6,124[a]	294[b]
			Year 2011			
1	Control	7.36[a]	27[a]	152[a]	7,136[a]	282[a]
2	N fertiliser	7.35[a]	26[a]	145[a]	7,093[a]	288[a]
3	Digestate	7.37[a]	46[b]	171[b]	7,191[a]	343[b]
4	NPKMg fertiliser	7.37[a]	40[b]	161[b]	7,214[a]	331[b]

Different letters (a, b) within the columns indicate significant differences between treatments for each year extra (P <0.05)

In the treatment with digestate and that with NPKMg fertiliser, the contents of P, K and Mg in the soil were found to have increased significantly after harvest compared with the other treatments (Table 5). Thus the soil was enriched with these nutrients after application of these two fertilisers despite the removal of nutrients in the harvested kohlrabi. Both treatments also produced considerably higher yields of bulbs (and hence also higher nutrient uptake) than the other treatments. Loria and Sawyer [21] reported that anaerobically digested liquid swine manure could provide similar plant-available N and P as would be expected from raw swine manure. In the present study, there were no differences between treatments in terms of soil pH (Table 5). Digestate is alkaline and increases the soil pH [22] making it suitable for acid soils. The reason for the reduction in soil K content in the N fertiliser treatment was its higher uptake by plants as a result of the synergistic effect of N fertilisation [23].

Nitrogen is the nutrient most subject to transformations affecting its availability to plants. These transformations include mineralisation, immobilisation, nitrification and denitrification, as well as leaching and ammonia volatilisation [24]. It is difficult to synchronise the supply of N from organic manures with the demand by crops for N [25]. Soil microbial activity leads to N release that is not necessarily in synchrony with plant nutrient demand [26]. The elevated NH_4^+-N concentration in digested effluents suggests its potential suitability as a readily available N fertiliser source.

The post-harvest soil content of N_{min} was lowest in the unfertilised treatment (Table 6). No appreciable differences in the total content of N_{min} (6.12-7.57 mg/kg and 4.02-4.70 mg/kg respectively) were observed among the other treatments. However, a difference was observed in the forms of N present. The proportion of N-NH_4^+ was greater than the N-NO_3^- only in the digestate treatment. Digestate has a large proportion of organically bound N (50-75%) which is available only after mineralisation [27]. The results suggest that, due to the short duration of the experiments, only a minor amount of this total N was nitrified. For the plants to utilise more N from the digestate, a longer period of crop growth and N uptake would be necessary.

Table 6. Content of mineral nitrogen (N_{min}) in the soil after harvest (2010 and 2011)

Treatment No.	Description	mg/kg DM		
		$N-NH_4^+$	$N-NO_3^-$	N_{min}
		Year 2010		
1	Control	1.56	1.71	3.27[a]
2	N fertiliser	2.45	5.12	7.57[b]
3	Digestate	3.80	2.32	6.12[b]
4	NPKMg fertiliser	3.08	4.10	7.18[b]
		Year 2011		
1	Control	1.11	1.30	2.41[a]
2	N fertiliser	1.73	2.29	4.02[b]
3	Digestate	2.73	1.97	4.70[b]
4	NPKMg fertiliser	1.85	2.35	4.20[b]

DM – dry matter; Different letters (a, b) within the columns indicate significant differences between treatments for each year extra ($P < 0.05$)

Weight of single bulbs

One characteristic of kohlrabi is a high demand for N from the soil [28], and therefore deficiency of NO_3-N in the soil reduces yields [29]. Sharof and Weir [30] studied the minimum amount of N required by vegetable crops, including kohlrabi, in relation to N balance in the soil and found that N requirements were invariably lower than values indicated by field trials.

As early as the first stages of growth in this pot study, there was a visible difference between the fertilised treatments and the unfertilised control. The plants in the latter had a lighter colour and growth of the above-ground biomass was markedly slower. At harvest, symptoms of P deficiency (violet discolouration) were detected on bulbs of the control treatment, which was the result of low P supply to the soil and relatively unsuitable pH value for P uptake.

The weight of the unfertilised bulbs was 48.5% (2010) and 66.2% (2011) lower than those in the treatment with N fertiliser only (Table 7). This demonstrates that N is an important element in terms of yield [31, 28]. However the weight of single bulbs fertilised with the digestate and with NPKMg fertilisers was significantly higher, by 34.7-42.9% (digestate) and 37.2-38.2% (NPKMg fertilisers) respectively, than that of bulbs fertilised with N fertiliser only. The positive synergistic effect of additional nutrients (especially P, K, Mg) on yield formation was apparent for the digestate and NPKMg fertiliser treatments. However, no significant differences were found between these two treatments. Lošák et al. [39] found the same yield results with kohlrabi of a different variety (Segura F1).

Table 7. Weight of single bulb for each of the treatments (2010 and 2011)

Treatment No.	Description	Weight of one bulb			
		Year 2010		Year 2011	
		g	rel. %	g	rel. %
1	Control	66[a]	51.5	40[a]	33.8
2	N fertiliser	128[b]	100.0	118[b]	100.0
3	Digestate	183[c]	142.9	159[c]	134.7
4	NPKMg fertiliser	177[c]	138.2	162[c]	137.2

Different letters (a, b, c) within the columns indicate significant differences between treatments for each year extra ($P < 0.05$)

In experiments lasting several years, Stinner et al. [3] also reported positive effects of three different types of digestate (fermented clover-grass mixture, cover crops and post-harvest residues) on wheat yields. Similarly, Bath and Elfstrand [7] reported that yields of leek were higher after digestate application than after fertilisation with compost. On soil with a low or satisfactory supply of available nutrients, Cigánek et al. [1] found that grain yield of winter wheat increased by 30.0-63.9% and seed yield of winter rape by 38.5–57.7% when fertilised with digestate compared with an unfertilised control.

Content of ascorbic acid and nitrates in bulbs

Vitamin C, including ascorbic acid and dehydroascorbic acid, is one of the important nutritional quality factors in many horticultural crops and has many biological functions in the human body. The content of vitamin C in vegetables can be influenced by various factors such as genotype differences, pre-harvest climatic conditions and cultural practices, maturity and harvesting method, and post-harvest handling procedures [32]. Mozafar [33] reported that N fertilisers, especially at high rates, seem to decrease the concentration of vitamin C in many different vegetables. In contrast, Nilsson [34] reported that N fertilisation did not affect the content of vitamin C in cauliflower.

The lowest content of ascorbic acid in kohlrabi in this study was observed in the unfertilised control (Table 8). A number of authors have concluded that adequate nutrition and fertilisation helps to increase yields and quality parameters, e.g. vitamin content [31, 23]. Maurya et al. [35] showed that with higher N doses cauliflower contained significantly more vitamin C. No significant differences in the ascorbic acid content were detected between the fertilised treatments (Table 8).

The concentration of NO_3^- in plants is affected primarily by the vegetable species, level of N fertilisation, plant organ analysed, growth stage and the sulphur concentration in the tissues [36, 37]. Kohlrabi is prone to a higher risk of nitrate accumulation in tissues than some other vegetables [31]. The lowest nitrate content was observed in the unfertilised treatment and the second lowest nitrate content in the digestate treatment (Table 8). The same results were described by Lošák et al. [39] with kohlrabi variety Segura F1. The reason is probably that the digestate contains a significant proportion of organic N (25-50%), which is subject to mineralisation and subsequent nitrification only after a certain period [27].

Table 8. Content of ascorbic acid and nitrates in kohlrabi bulbs grown in the different treatments

Treatment No.	Description	Content of ascorbic acid (mg/kg FM)		Nitrate content (mg/kg FM)	
		Year 2010	Year 2011	Year 2010	Year 2011
1	Control	552[a]	311[a]	36[a]	58[a]
2	N fertiliser	763[b]	354[b]	798[c]	695[c]
3	Digestate	774[b]	364[b]	197[b]	230[b]
4	NPKMg fertiliser	768[b]	352[b]	785[c]	699[c]

FM – fresh matter; Different letters (a, b, c) within the columns indicate significant differences between treatments for each year extra (P <0.05)

It can be assumed that during the short growing period of kohlrabi (approx. 7 weeks), only part of the organically bound nitrogen was mineralised. Therefore mineral $N-NH_4^+$ from the digestate (or after its nitrification $N-NO_3^-$) was available to the plants and was sufficient for yield formation, but did not increase the nitrate content in the bulbs. The nitrate content was highest in the two treatments fertilised with nitrogen in the form of urea (N and NPKMg fertiliser treatments), where it was threefold that in the digestate treatment (Table 8).

Macronutrients content of the bulbs

In terms of the content of macronutrients in the bulbs (Table 9), the differences were most marked in the case of nitrogen. The N content was highest in the urea-fertilised treatments (N and NPKMg fertiliser treatments) from which the plants could exploit the available forms, i.e. NH_4^+ and NO_3^-.

In the digestate treatment, the N content in kohlrabi tissues was significantly lower as a result of limited mineralisation of organically bound N within the short time interval. Furthermore the addition of readily degradable C compounds may have led to immobilisation of mineral N in soil [26]. In terms of the other nutrients, there were mostly no significant differences between the treatments.

Table 9. Contents of macronutrients in the bulbs grown using the different treatments

Treat. No.	Description	% in DM				
		N	P	K	Ca	Mg
		Year 2010				
1	Control	1.16^a	0.28^b	2.68^a	0.25^a	0.12^a
2	N fertiliser	2.72^c	0.20^a	2.72^a	0.24^a	0.13^a
3	Digestate	1.61^b	0.27^b	2.72^a	0.25^a	0.11^a
4	NPKMg fertiliser	2.23^c	0.28^b	2.80^a	0.27^a	0.12^a
		Year 2011				
1	Control	1.32^a	0.25^{ab}	3.01^a	0.21^a	0.14^a
2	N fertiliser	2.55^c	0.22^a	2.99^a	0.23^b	0.13^a
3	Digestate	2.16^b	0.29^c	3.11^a	0.23^b	0.13^a
4	NPKMg fertiliser	2.67^c	0.28^{bc}	3.21^a	0.24^b	0.15^a

DM – dry matter; Different letters (a, b, c) within the columns indicate significant differences between treatments for each year extra (P <0.05)

A low P content was detected in the tissues of plants where urea alone was applied (treatment 2), which was due to its low content in the soil and alkaline pH negatively affecting its uptake [23]. Möller and Stinner [38] reported that slurry digestion did not influence plant P and K uptake.

CONCLUSIONS

Digestate application resulted in comparable or better soil chemical properties, crop yield and quality parameters of kohlrabi than mineral fertiliser application. When available, the application of digestate can potentially offer considerable savings compared with the purchase of mineral fertilisers. However, digestates are poor in labile organic substances and the soil must be supplied with these from other sources, if required.

ACKNOWLEDGEMENT

This study was supported by the Internal Grant Agency of the Faculty of Agronomy of Mendel University in Brno No. TP 5/2010, TP 9/2011 and TP 11/2013. We thank to Mr. Chris Dawson from the UK for the language revision.

REFERENCES

1. Cigánek, K., Lošák, T., Szostková, M., Zatloukalová, A., Pavlíková, D., Vítěz, T., Fryč, J., Dostál, J., Verification of the Effectiveness of Fertilisation with Digestates from Biogas Stations on Yields of Winter Rape and Winter Wheat and changes in some Agrochemical Soil Properties, *Agrochemistry*, Vol. 50, No. 3, pp 16-21, 2010.
2. Möller, K. and Müller, T., Effect of Anaerobic digestion on Digestate Nutrient availability and Crop growth: A Review, *Engineering in Life Sciences*, Vol. 12, No. 3, pp 242-257, 2012.
3. Stinner, W., Möller, K. and Leithold, G., Effects of Biogas digestion of Clover/grass-leys, Cover Crops and Crop Residues on Nitrogen Cycle and Crop Yield in Organic stockless Farming Systems, *European Journal of Agronomy*, Vol. 29, No. 2-3, pp 125-134, 2008.
4. Arthurson, V., Closing the Global Energy and Nutrient Cycles through Application of Biogas Residue to Agricultural Land-potential Benefits and Drawbacks, *Energies*, Vol. 2, No. 2, pp 226-242, 2009.
5. Gunnarsson, A., Bengtsson, F. and Caspersen, S., Use efficiency of Nitrogen from Biodigested Plant Material by Ryegrass, *Journal of Plant Nutrition and Soil Science*, Vol. 173, No. 1, pp 113-119, 2010.
6. Ross, D. J., Tate, K. R., Speir, T. W., Stewart, D. J. and Hewitt, A. E., Influence of Biogas-digester effluent on Crop growth and Soil Biochemical Properties under Rotational cropping, *New Zealand Journal of Crop and Horticultural Science*, Vol.17, No.1, pp 77-87, 1989.
7. Bath, B. and Elfstrand, S., Use of Red Clover-based Green Manure in Leek Cultivation, *Biological Agriculture & Horticulture*, Vol. 25, No. 3, pp 269-286, 2008.

8. Odlare, M., Pell, M. and Svensson, K., Changes in Soil Chemical and Microbiological Properties during 4 years of Application of Various Organic Residues, *Waste Management*, Vol. 28, No. 7, pp 1246-1253, 2008,

9. Kolář, L., Kužel, S., Peterka, J. and Borová-Batt, J., Agrochemical Value of the Liquid Phase of Wastes from Fermenters during Biogas Production, *Plant, Soil and Environment*, Vol. 56, No. 1, pp 23-27, 2010.
10. Lošák, T., Zatloukalová, A., Szostková, M., Hlušek, J., Fryč, J., Vítěz, T., Comparison of the Effectiveness of Digestate and Mineral Fertilisers on Yields and Quality of Kohlrabi (Brassica oleracea, L.), *Acta Universitatis Agriculturae et Silviculturae Mendeleianae Brunensis*, Vol. 59, No. 3, pp 117-122, 2011.

11. Möller, K., Influence of different Manuring Systems with and without Biogas Digestion on Soil Organic Matter and Nitrogen Inputs, Flows and Budget in Organic Cropping Systems, *Nutrient Cycling in Agroecosystems*, Vol. 84, No. 2, pp 179-202, 2009.
12. Amon, T. and Döhler, H., *Handreichung-Biogasgewinnung und –nutzung,* Fachagentur Nachwachsende Rohstoffe e.V. (ed.), KTBL, Darmstadt, 2004.
13. Quakernack, R., Pacholski, A., Techow, A., Herrmann, A., Taube, F. and Kage, H., Ammonia volatilization and Yield response of Energy Crops after fertilization with Biogas Residues in a Coastal marsh of Northern Germany, *Agriculture, Ecosystems and Environment,* Vol. 160, 2011.
14. Gericke, D., Bornemann, L., Kage, H. and Pacholski, A., Modelling Ammonia losses after Field application of Biogas Slurry in Energy Crop Rotations, *Water Air and Soil*

Pollution,   Vol.   223,   No.   1,   pp   29-47,   2012.

15. Alburquerque, J. A., de la Fuente, C. and Bernal, M. P., Chemical properties of Anaerobic Digestates affecting C and N dynamics in Amended Soils, *Agriculture, Ecosystems   and   Environment,*   Vol.   160,   2011.

16. Möller, K., Schulz, R. and Müller, T., Effects of setup of Centralized Biogas Plants on Crop Acreage and Balances of Nutrients and Soil Humus, *Nutrient Cycling in Agroecosystems,*   Vol.   89,   No.   1,   pp   303-312,   2011.

17. Zbíral, J., *Soil Analysis I, Integrated Work Procedures,* ÚKZÚZ, Brno, 2002.

18. Zbíral, J., *Plant Analysis, Integrated Work Procedures,* ÚKZÚZ, Brno, 1994.

19. Plaixats, J., Barcelo, J. and Garcia-Moreno, J., Characterization of the Effluent Residue from Anaerobic digestion of Pig Excreta for its Utilization as Fertilizer, *Agrochemica,* Vol. 32, No. 2-3, pp 236-239, 1988.

20. Gerardi, M. H., *The Microbiology of Anaerobic Digesters,* John Wiley and Sons, Inc: Hoboken, NJ, U.S.A., 2003.

21. Loria, E. R. and Sawyer, J. E., Extractable Soil Phosphorus and Inorganic Nitrogen following Application of Raw and Anaerobically digested Swine Manure, *Agronomy Journal,* Vol. 97, No. 3, pp 879-885, 2005.

22. Fuchs, J. G., Berner, A., Mayer, J. and Schleiss, K., Influence of Composts and Digestats on Soil Fertility, *Agrarforschung,* Vol. 15, No. 6, pp 276-281, 2008.

23. Mengel, K. and Kirkby, E. A., *Principles of Plant Nutrition,* 5th edition, Kluwer Academic   Publishers,   Dordrecht/Boston/London,   2001.

24. Möller, K. and Stinner, W., Influence of different Manuring Systems with and without Biogas digestion on Soil Mineral Nitrogen Content and on Gaseous Nitrogen losses (ammonia, nitrous oxides), *European Journal of Agronomy,* Vol. 30, No. 1, pp 1-16, 2009.

25. Pang, X. P. and Letey, J., Organic farming: Challenge of timing Nitrogen availability to crop Nitrogen Requirements, *Soil Science Society of America Journal,* Vol. 64, No. 1, pp 247-253, 2000.

26. Dosch, P. and Gutser, R., Reducing N losses (NH_3, N_2O, N_2) and Immobilization from Slurry through Optimized Application Techniques, *Fertilizer Research,* Vol. 43, No. 1-3, pp 165-171, 1996.

27. Kirchmann, H. and Witter, E., Composition of Fresh Aerobic and Anaerobic Farm Animal Dungs, *Bioresource Technology,* Vol. 40, No. 2, pp 137-142, 1992.

28. Feller, C. and Fink, M., Nitrogen uptake of Kohlrabi, estimated by Growth Stages and an Empirical Growth Model, *Journal of Plant Nutrition and Soil Science,* Vol. 160, No. 6, pp 589-594, 1997.

29. Steingrobe, G. and Schenk, M. K., Influence of Nitrate Concentration at the Root Surface on Yield and Nitrate uptake of Kohlrabi (Brassica oleracea – gongyloides, L.) and Spinach (Spinacia oleracea, L.), *Plant and Soil,* Vol. 135, No. 2, pp 205-211, 1991.

30. Sharof, H. C. and Weir, U., *Calculation of nitrogen immobilization and fixation,* Gartenbau Hannover Germany, Bodenkunde, 1994.

31. Hlušek, J., Richter, R. and Ryant, P., *Výživa a hnojení zahradních plodin,* 1th ed., Zemědělec, Praha, 2002. (in czech)

32. Lee, S. K. and Kader, A. A., Preharvest and Postharvest Factors influencing Vitamin C Content of Horticultural Crops, *Postharvest Biology and Technology*, Vol. 20, No. 3, pp 207-220, 2000.

33. Mozafar, A., Nitrogen Fertilizers and the amount of Vitamins in Plants: A Review, *Journal of Plant Nutrition*, Vol. 16, No. 12, pp 2479-2506, 1993.

34. Nilsson, T., The influence of Soil Type, Nitrogen and Irrigation on Yield, Quality and Chemical Composition of Cauliflower, *Swedish Journal of Agricultural Research*, Vol. 10, No. 2, pp 65-75, 1980.

35. Maurya, A. N., Chaurasia, S. N. S. and Reddy, Y. R. M., Effect of Nitrogen and Molybdenum Levels on Growth, Yield and Quality of Cauliflower (Brassica oleracea var. Botrytis) cv. Snowball-16, *Haryana Journal of Horticultural Sciences*, Vol. 21, No. 3-4, pp 232-235, 1992.

36. Lošák, T., Hlušek, J., Kráčmar, S. and Varga, L., The effect of Nitrogen and Sulphur Fertilization on Yield and Quality of Kohlrabi (Brassica oleracea, L.), *Revista Brasileira de Ciencia do Solo*, Vol. 32, No. 2, pp 697-703, 2008.

37. Marschner, H., *Mineral Nutrition of Higher Plants,* 2nd edition, Academic Press, London, 2002.

38. Möller, K. and Stinner, W., Effect of Organic Wastes digestion for Biogas production on Mineral Nutrient availability of Biogas Effluents, *Nutrient Cycling in Agroecosystems*, Vol. 87, No. 3, pp 395-413, 2010.

39. Lošák, T., Musilová, L., Zatloukalová, A., Szostková, M., Hlušek, J., Fryč, J., Vítěz, T., Haitl, M., Bennewitz, E. and Martensson, A., Digestate is Equal or a Better Alternative to Mineral Fertilization of Kohlrabi, *Acta Universitatis Agriculturae et Silviculturae Mendeleianae Brunensis*, Vol. 60, No. 1, pp 91-96, 2012.

14

Dynamic Modeling of Kosovo's Electricity Supply–Demand, Gaseous Emissions and Air Pollution

Sadik Bekteshi[*1], *Skender Kabashi*[2], *Skender Ahmetaj*[3], *Ivo Šlaus*[4],
Aleksander Zidanšek[5], *Kushtrim Podrimqaku*[6], *Shkurta Kastrati*[7]

[1]Faculty of Natural Sciences, University of Prishtina "Hasan Prishtina", Prishtina, Kosovo
e-mail: sadik.bekteshi@uni-pr.edu
[2]Faculty of Natural Sciences, University of Prishtina "Hasan Prishtina", Prishtina, Kosovo
e-mail: skenderkabashi@yahoo.com
[3]Faculty of Natural Sciences, University of Prishtina "Hasan Prishtina", Prishtina, Kosovo
e-mail: skenderahmetaj@yahoo.com
[4]Ruđer Bošković Institute, Bijenička 54, Zagreb, Croatia
e-mail: slaus@irb.hr
[5]Jožef Stefan International Postgraduate School, Faculty of Natural Sciences and Mathematics, University
of Maribor, Jamova 39, Ljubljana, Slovenia
e-mail: Aleksander.Zidansek@ijs.si
[6]Faculty of Electrical and Computer Engineering, University of Prishtina, "Hasan Prishtina", Mother
Teresa Str., Prishtina, Kosovo
e-mail: kushtrim.podrimqaku@uni-pr.edu
[7]Department of Power Engineering Equipment, Technická univerzita, Studentská 1402/2, Liberec, Czech
Republic
e-mail: kastratishkurta@hotmail.com

ABSTRACT

In this paper is described the developing of an integrated electricity supply–demand, gaseous emission and air pollution model for study of possible baseline electricity developments and available options to mitigate emissions. This model is constructed in STELLA software, which makes use of Systems Dynamics Modeling as the methodology. Several baseline scenarios have been developed from this model and a set of options of possible developments of Kosovo's Electricity Supply–Demand and Gaseous Emissions are investigated. The analysis of various scenarios results in Medium Growth Scenarios (MGS) that imply building of generation capacities and increase in participation of the electricity generation from renewable sources. MGS would be 10% of the total electricity generation and ensure sustainable development of the electricity sector. At the same time, by implementation of new technologies, this would be accompanied by reduced Greenhouse Gases (GHG) (CO_2 and NO_x) emissions by 60% and significant reduction for air pollutants (dust and SO_2) by 40% compared to the Business-As-Usual (BAU) case. Conclusively, obtained results show that building of new generation capacities by introducing new technologies and orientation on environmentally friendly energy sources can ensure sustainable development of the electricity sector in Kosovo.

KEYWORDS

Modeling, Electricity demand, Electricity supply, Emissions, Scenarios, System dynamics.

INTRODUCTION

Energy, with a heavy reliance on fossil fuels, is closely intertwined with climate change and sustainable development. While climate change is obviously a global

---

[*] Corresponding author

environmental problem, nevertheless there is potential for local actions for sustainable transition to renewable locally produced energy and CO_2 and air pollutants emission reduction.

Nowadays, there are numerous models that can generate possible scenarios for the evolution of energy systems applying different approaches such as:

- A medium- to long-term energy system planning;
- Energy policy analysis, and scenario development [1];
- The different potential energy mixes not only for the whole of Europe, but also for defined regions or countries [2];
- Long-term projections of energy supply, demand, and prices by sector [3];
- Electricity prices and fuel costs [4];
- Integration of renewable sources in the energy systems of island or other isolated locations [5], etc.

Connolly *et al.* [6] give a recent review of computer tools for analyzing the integration of renewable energy into various energy systems.

The objective of this paper is to develop a dynamic electricity demand-supply and gaseous emission model that could be used for these main purposes:

- To examine mid- to long-term scenario studies simulating the relevant developments of electricity supply-demand for the Republic of Kosovo;
- To simulate emissions of CO_2, SO_2, NO_x and dust from coal combustion in power plants. The projections can be evaluated to determine the ability of various reduction strategies to achieve desired CO_2, SO_2, NO_x and dust emission levels.

In this study, computer programming software, STELLA [7] was used to construct the dynamic model for electricity supply-demand and gaseous emissions system.

MODELING AND SIMULATION

In this paper is described the construction of a new model, which is referred to as a dynamic supply-demand and gaseous emission model. This model makes use of Systems Dynamics Modeling, developed by Forrester as a methodology [8], initially applied to corporate problems and called as industrial dynamics.

The issue of present and future demand and supply of electricity is of great interest for many aspects of the modern world. There are different sources of electricity demand which have different growth characteristics. Growth in demand sourced from the industry, for example, has different drivers to growth associated with domestic residences. Thus, one way of looking at the future of electricity demand and supply is to consider various trends in demand and supply at the end-user sector level. In the end-use approach the electricity demand sector is divided into relevant (sub) sectors (e.g., industry, transport, residential), which are then divided over the relevant services (e.g., space heating, lighting).

Electricity demand modeling considered in this study is focused on major electricity demand sectors:

- Residential sector;
  - Space heating;
  - Cooking;
  - Hot water and other uses.
- Industrial sector: including low and high voltage industrial customers;
- Service sector: including all low-voltage non-residential customers such as;
  - Commercial services;
  - Public services;
  - Agriculture, etc.

Electricity generation sectors;
- Nonrenewable energy sector:
  - o Electricity generation from nonrenewable sources.
- Renewable energy sector;
  - o Electricity generation from renewable sources, and for developing projections of CO_2, SO_2 and NO_x emissions from electricity generation.
- Gaseous and air pollution emission sector.

The sum of flows of electricity from electricity generation sectors is subtracted from the sum of electricity demands, giving the result for total electricity demand - supply balance[†] (see Appendix).

Residential sector

Accurate and reliable electricity demand forecasts are very important in order to develop alternatives and solutions for future electricity needs. This is particularly important for residential sector, since in many countries, it is a substantial consumer of electricity and therefore a focus for electricity consumption efforts. Electricity use in this sector includes energy for space heating, air conditioning, cooking, cleaning, washing, drying, lighting, cooling and another electrical uses.

Population is one of the most important factors that influence the electricity demand supply in residential sector. The size of population is depending on the population growth rate. Household size is included in the model to capture the impact of the number of inhabitants per household on the demand for electricity. The number of household is determined by two factors, i.e. household size (number of family member within a household) and total population. The total number of household has been calculated as the ratio between the population and the number of principal houses. The total electricity consumption by residential sector was obtained by multiplying the number of households with the average demand per household.

The total electricity consumption from the residential sub-model (see Appendix), was assumed to be based on four main factors i.e. population (*Population*), number of households (*Number_of_Households*),[‡] average electricity consumption per household (*Avg_demand_per_ Household_user*) and household electricity consumption growth rate (*Growth_rate_ demand_per_residential_user*). Average electricity consumption per household is calculated from the expression:

$$Avg\_demand\_per\_Household\_user(t) = Avg\_demand\_per\_Household\_user(t - dt) \\ + (Demand\_per\_residential\_user\_change) \times dt \tag{1}$$

where the variable (*Demand_per_residential_user_change*) represents the change of consumption of electricity during time *dt*. While modeling of electricity consumption in this sector, one should be aware that average electricity demand per household decreases slowly with time. For example, for US residential sector it was 12.3 MWh in 2011 and is expected to drop to 11.5 MWh in 2040 [6]. Thus, the consumption of the electricity from the residential sector is calculated by the expression:

$$Residential\_demand = Avg\_demand\_per\_Household\_user \times \\ Number\_of\_Households \tag{2}$$

[†] For details of the model contact the corresponding author.

[‡] Hereby, in order that the model be more readable, variables of the model will be labeled in a specific way: e.g., number of households as (*Number_of_Households*)

Industrial sector

Electricity is one of the basic inputs for industrial growth with significant impact on the overall performance of the national economy. Electricity is consumed in the industrial sector for a wide range of activities which are involved in manufacturing industries (food, pulp and paper, chemicals, refining, iron and steel, nonferrous metals, and nonmetallic minerals, among others) and in nonmanufacturing industries (agriculture, mining and construction, and non-specified others). Industrial electricity demand varies across regions and countries, depending on the level and mix of economic activity and technological development, among other factors.

During last few years, a tendency is observed toward an increase in the industrial sector. About 40% of industrial demand growth is driven by electricity in 2010 [9]. The trend of the industrial electricity consumption of a country is estimated by analyzing macro-economic indicators, Gross Domestic Product (GDP) trends and other factors that influence electricity demand of this sector [10]. Industry specific studies may yield information on expectations of growth within this sector [6].

Industrial electricity consumption is divided into two groups: distribution customer, supplied at Low and Medium Voltage (LV), and direct customers, supplied at High Voltage (HV). The basic calculations for estimation of electricity consumption by industrial sector are as follows: Electricity demand in the industrial sector (*Industrial_demand*) in this model is expressed as the sum of low voltage (*LV_industrial_demand*) and high industrial consumers (*HV_industrial_demand*). The dynamics of low voltage industrial sector demand depends on the number of industrial customers (*NLV_Industrial_user*) and is expressed by the equation:

$$NLV\_Industrial\_user(t) = NLV\_Industrial\_user(t - dt) + \\ (New\_LV\_Industrial\_user) \times dt \tag{3}$$

On the other hand, industrial growth is expected to raise the number of industrial customers (*New_LV_industrial_user*) and increase of electricity consumption by each industrial customer. Similar relations also apply to high voltage industrial sector.

Service sector

In the service sector, also referred to as the commercial or tertiary sector, are several groups of customers supplied at low voltage: artisans, commercial services, public services, public lighting, agriculture, public transport infrastructure and others. Knowledge about the electricity demand of the service sector by end-use or by sub-sector is crucial to determine the impact of novel technologies and other energy-efficiency measures. The methodologies used for estimating electricity consumption by service and industrial sector are similar.

Electricity demand in the service sector (*Service_Demand*) depends on the number of customers (*Number_of_Service_users*):

$$Service\_demand = Number\_of\_Service\_users \times Demand\_per\_Service\_user \tag{4}$$

and each customer demand for electricity (*Demand_per_Service_user*). On the other hand due to the growth of the service sector it is expected to raise the number of customers:

$$Number\_of\_Service\_users(t) = Number\_of\_Service\_users(t - dt) + \\ (New\_Service\_users) \times dt \tag{5}$$

and to increase electricity consumption by each consumer in this sector (*New_Service_users*).

Electricity generation sector

The electricity generation sector consists of two sub-sectors: nonrenewable energy sub-sector (that includes electricity generation from nonrenewable energy sources) and renewable energy sub-sector (that includes electricity generation from renewable energy sources). The total annual electricity generation E from the all production power plants and all renewable sources is calculated by the equation:

$$E = \sum_i NRE_i + RE_i \tag{6}$$

where:
- NRE_i - total annual electricity generation from nonrenewable electricity sources;
- RE_i - total annual electricity generation from renewable electricity sources respectively, for the given year (*i*).

Gaseous emission sector

The continuing growth of the world energy consumption during recent decades has also influenced its level of carbon dioxide (CO_2), sulfur dioxide (SO_2) and nitrogen oxides (NO_x) emissions. Since the Industrial Revolution, annual global carbon dioxide emissions from fossil-fuel combustion dramatically increased to over 31 Gt CO_2 in 2011 [11].

Fuel combustion in power plants for electricity generation is a complex phenomenon which results in emissions of different gases at different levels. There are several types of fuel that can be used by power plants: traditional fuels (coal, lignite, peat, natural gas, oil, and uranium) and also: Municipal Solid Waste (MSW), biofuel and hydrogen or blast furnace gas.

The projection of CO_2, SO_2, NO_x and dust emissions from combustion in stationary plants is also estimated by this model. The gaseous emissions per unit per year can be calculated from the electricity generated, the efficiency and specific gaseous emissions of various fuels (for lignite, see Table.1). If fuels are used for electricity generation, gaseous emissions increase with the reciprocal of the power plant efficiency. The summation of these individual gas emission data leads to the total annual emission of GHG and air pollutants E_i per unit i of power generation from Thermal Power Plant (TPP) for a specific fuel used according to the formula:

$$E_i = \sum \frac{EG_i \times EF_{ki}}{Ef_i} \tag{7}$$

where:
- $k = CO_2$, NO_x, SO_2 stands for respective gases;
- EG_i - electricity generated per unit i [kWh];
- Ef_i - power plant efficiency;
- EF_{ki} - specific gaseous emissions of various fuels (kg of specific gaseous emission per MWh).

Initial values of the emission factors are taken from the values presented in Table 1.

Energy, economy and environment models are often used to look at opportunities and costs of reducing Greenhouse Gases (GHGs). Achieving this reduction using energy from biomass, solar, wind and other renewable sources and/or nuclear energy seems to be a viable alternative [13]. Otherwise, the response of the climate system to increasing greenhouse gas concentrations and other anthropogenic forces includes the potential for 'rapid climate change' [14].

Table 1. Emission factors for lignite for Kosova power plants: unit A (EF_A); unit B (EF_B); and New Kosova (EF_C) [12]

CO_2 [t/MWh]			NO_x [kg/MWh]			SO_2 [kg/MWh]			Dust [kg/MWh]			CH_4 [kg/MWh]			CO [kg/MWh]		
EF_A	EF_B	EF_C	EF_A	EF_B	EF_C	EF_A	EF_B	EF_C	EF_A	EF_B	EF_C	EF_A	EF_B	EF_C	EF_A	EF_B	EF_C
1.5	1.4	0.8	4	3.8	1.55	3.2	3.1	0.5	3.4	1.4	0.05	0.33	0.3	0.07	0.065	0.06	0.027

Electricity demand for all sectors

The total net electricity demand is the sum of the electricity demands for all three sectors, residential, service and industrial, for each year. However, all electricity supplied to a distribution utility does not reach the end consumer, a portion of electricity lost. Electricity losses can be divided into two categories: technical and non-technical (or commercial). Technical losses occur naturally and consist mainly of power dissipation in electricity system components such as transmission and distribution lines, transformers, and measurement systems. There are many types of non-technical losses: non-payment by customers, electricity theft, errors in technical losses computation, errors in accounting and record keeping that distort technical information, etc. Electricity losses are unavoidable and must be modeled before accurate representation can be calculated. Electricity demand from all three sectors is calculated from the expression:

$$Total\_net\_demand = \qquad\qquad (8)$$
$$Industrial\_demand + Residential\_demand + Service\_demand$$

Total electricity demand (annual average) is the total net demand of customers increased by transmission and distribution losses that must be supplied to the electrical system by the generation plants and/or by importing from neighboring countries.

One of the main factors that influence the electricity demand is the population. In Figure 1a, a gradual increase of population and number of households is seen. It must be stressed that, as far as we know, these are among the first published results in connection with the case study, which use the data of Population Registration Act of 2011 [15], which is 1,789 million. Since the previous registration had been carried out 30 years ago, most publications so far relied on numbers differing significantly from those in [12, 16], a fact that influenced the obtained results.

Up to 95-97% of the electricity in Kosovo is provided by the lignite power plants: Kosova A and Kosova B with effective capacity 870 MW. The construction of the TPP New Kosova is planned for the year 2018 [16]. In the same year TPP Kosova A is planned to go out of service. The remaining 3-5% of electric power only is produced from renewable resources: hydroelectric power generation is provided mainly by the Hydro Power Plant (HPP) Ujman with a capacity of 2×17.5 MW = 35 MW and HPP of Lumbardhi with a capacity of 8.3 MW. Thus, production capacity at disposal is about 900 MW. Total annual electricity generation from all production units is calculated by eq. (6). Thus, electricity consumption for 2013 was 4,944 GWh, out of which 2,852 GWh was used for residential, 1,285 GWh for industry and 807 GWh for services and it is foreseen that this electricity consumption will be gradually increased according to Figure 1b [16]. The total net demand forecasts for users are calculated by summing the electricity demands for all three sectors, for each year.

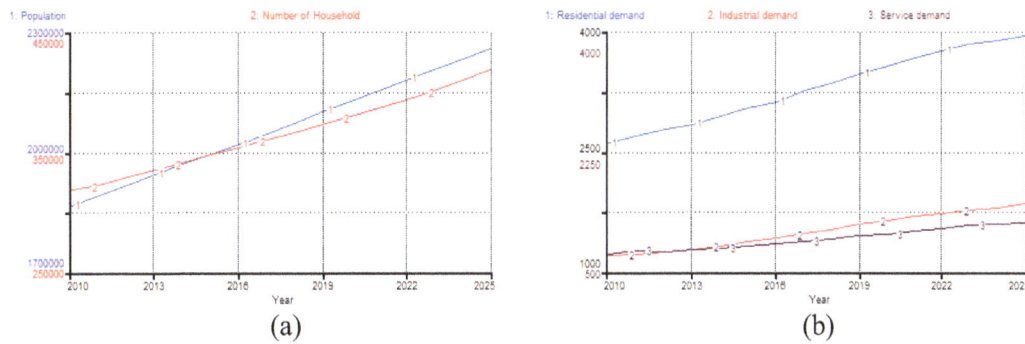

(a)   (b)

Figure 1. Population growth (blue line - 1), and total number of households (red line - 2) (a); adapted from [15, 17]. Forecast of electricity demand in the residential [GWh] (blue line - 1), industrial sector [GWh] (red line - 2) and service sector [GWh] (black line - 3) (b); adapted from [17]

Figures 2a and 2b show the future development of Kosovo's power system without any additional energy or climate policies implemented (the-Business-As-Usual, BAU, scenario). During the whole period 1999-2010, annual domestic electricity generation has been below the level of demand. This shortage of electricity will continue increasing if new generating capacities are not built (Figure 2a) in the near future, which is accompanied by an increase of greenhouse gas emissions and air pollution (Figure 2b). From this model, three baseline scenarios are simulated for new nonrenewable generation capacity (300 MW, 600 MW and 1,000 MW). For the same period, projected electricity generation from renewable resources is simulated for three scenarios, with an average growth of approximately 10% of the total electricity generation [18].

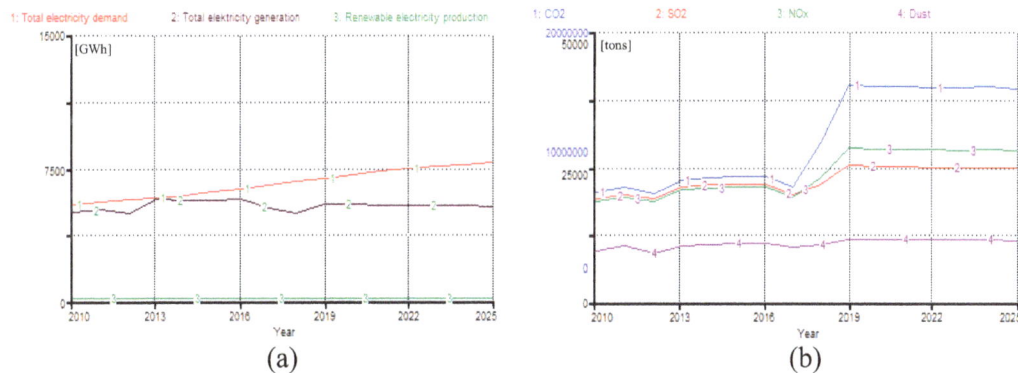

(a)   (b)

Figure 2. Values obtained from the model: total electricity demand (red line - 1), total electricity generation (black line - 2) and renewable electricity generation (green line - 3) (a); and CO_2 (blue line - 1), SO_2 (red line - 2), NO_x (green line - 3) and dust (pink line - 4) emissions (b); for Business-As-Usual (BAU) case

Low Growth Scenarios (LGS). According to this scenario, construction of a TPP New Kosova with 300 MW is taken into account, while the percentage of the participation of renewable energy in the total electricity generation is 10%. From Figure 3a it is seen that the values of total electricity generation (black line-2) are always less than the values of electricity demand (red line-1). This means that consumers' demands for electricity would not be fulfilled, so that Kosovo's electric power system would not be self-sufficient; therefore we will not proceed with disussion of this scenario. According to this scenario, emission of gases and dust are still at low levels (in comparison with other scenarios) and after the year 2018, indicate a slight and stable decrease until the year 2025 (Figure 3b).

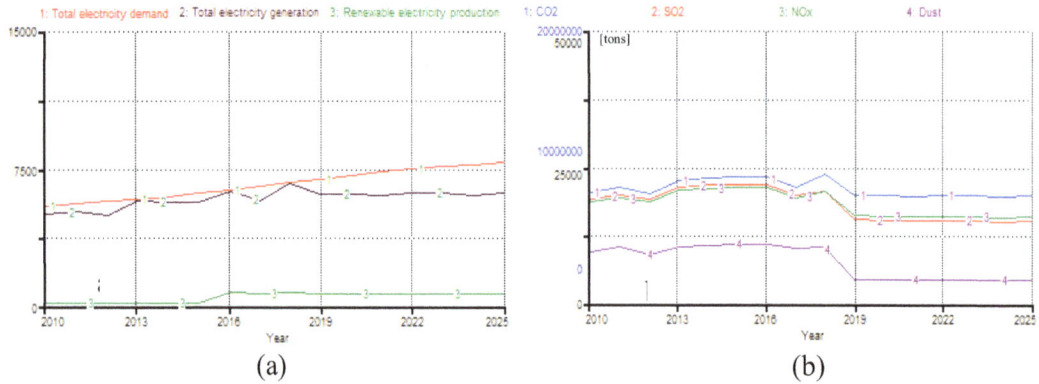

(a)   (b)

Figure 3. Values obtained from the model: total electricity demand (red line - 1), total electricity generation (black line - 2) and renewable electricity generation (green line - 3) (a); and CO_2 (blue line - 1), SO_2 (red line - 2), NO_x (green line - 3) and dust (pink line - 4) emissions (b); for LGS scenario

<u>Medium Growth Scenarios (MGS)</u>. According to this scenario, apart from electric power produced from the existing lignite power plants, construction of a TPP New Kosova with 600 MW is taken into account, while the renewable energy from the year 2018 to the year 2025 will consist of 10% of the total electricity generation in Kosovo.

In Figure 4a, this is presented as a rapid increase in the black line, which represents the total electricity generation. This curve is very close to the red-line curve (total electricity demand), which means that, to have a sustainable development of the energetic system the increase of the renewable energy must be 10%. From Figure 4b, we see that from the year 2010 until the year 2013, as a result of increase in electricity consumed, we have an almost linear increase of emission of gases and dust. Thereafter, stabilization and slight increase of GHG gas emissions and air pollution occur. In the long term, from year 2017 to 2025 we will have a significant reduction of NO_x and air pollutants (dust and SO_2) while values of CO_2 show a very slight increase in comparison with previous values.

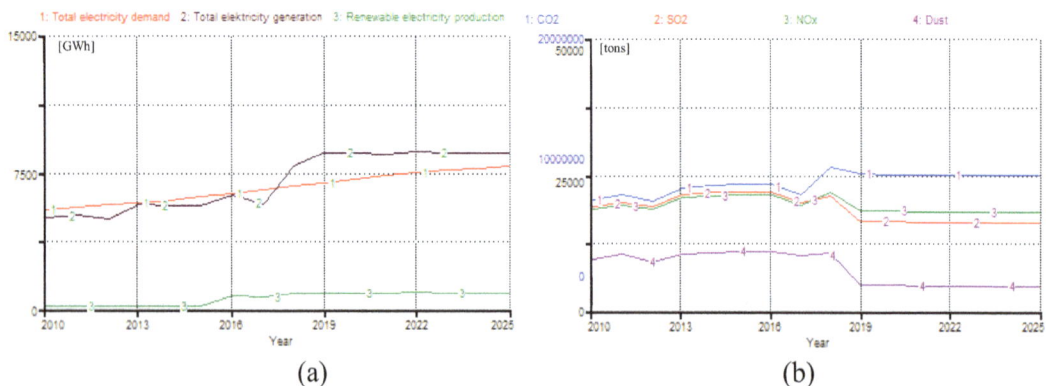

(a)   (b)

Figure 4. Values obtained from the model: total electricity demand (red line - 1), total electricity generation (black line - 2) and renewable electricity generation (green line - 3) (a); CO_2 (blue line - 1), SO_2 (red line - 2), NO_x (green line - 3) and dust (pink line - 4) emissions (b); for MGS scenario

Compared to the Business-As-Usual (BAU) case, CO_2 emission is reduced by 60% and air pollutants (dust and SO_2) and NO_x are significantly reduced by 40%. These reductions would be the consequence of the use of renewable energy resources (10% of the total electricity generation) and of new technologies in generation of electricity from lignite. Considering resources of renewable and nonrenewable energy of Kosovo [17], such a scenario (building of a TPP with a capacity of 600 MW and with electricity generation from renewable energy resources of 800 GWh) could be realized, and would provide an optimal solution in fulfillment of demands for electricity supply and environmental demands as well.

High Growth Scenarios (HGS). Basic considerations of this scenario is that the Kosovo's electric power system will advance mainly due to the high increase of electricity production from lignite. Apart from electric power produced from the lignite power plants Kosova A (TPP Kosovo A will remain in service through 2018) and Kosova B, in this scenario the growth of electricity generation from the TPP New Kosova with a capacity 1,000 MW, planned to start operating at 2018 is also considered, as well as the participation of renewable energy with 10% of the total electricity generation.

In Figure 5a, on account of a significant increase in total electricity generation, a large discrepancy is observed between it and the total electricity demand. In Figure 5b, it is shown that according to this scenario, there is a rapid increase of CO_2 after the year 2019. Taking into consideration the high cost of building such capacities as well as the impact on the environment that could be considerable since CO_2 could increase sharply, it seems that this scenario is not an optimal solution.

From analysis of these scenarios it follows that the most favorable and most realizable scenario is the scenario MGS, i.e. building of TPP New Kosova with capacity 600 MW and of capacities with renewable energy with at least 10% of the total electricity generation.

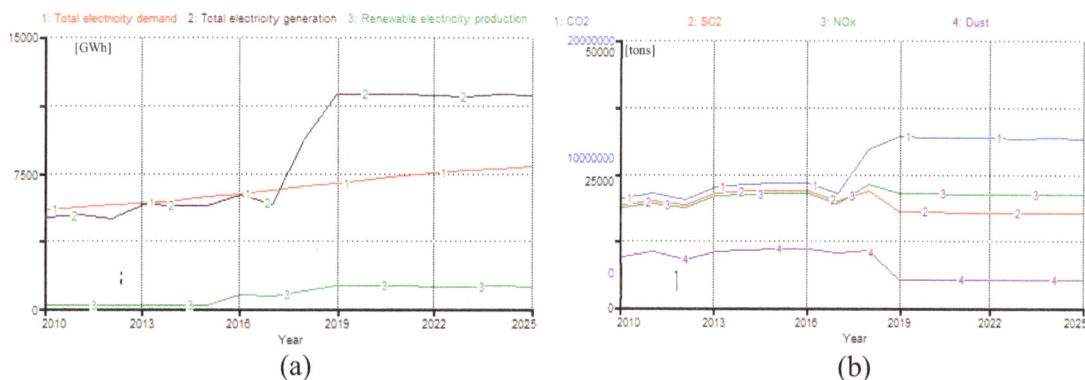

Figure 5. Values obtained from the model: total electricity demand [GWh] (red line - 1), total electricity generation [GWh] (black line - 2) and renewable electricity generation [GWh] (green line - 3) (a); CO_2 (blue line - 1), SO_2 (red line - 2), NO_x (green line - 3) and dust (pink line - 4) emissions (b); for HGS scenario

CONCLUSIONS

A new model has been developed for analyzing of the current situation of electricity generation and consumption and recent trends electricity development of Kosovo, as well. It is used to also develop projections of CO_2, SO_2, NO_x and dust emissions from electricity generation. The time span used here is the period 2010 – 2025. From the results obtained, it can be concluded that:

- Due to the population increase in Kosovo during the period 2010 – 2025, our scenarios indicate a strong increase in electricity use in all investigated sectors, and in the demand for electricity in general;
- According to HGS (building of a TPP with a capacity of 1,000 MW), Kosovo's electric power system could fulfill completely and even exceed demands for

electricity supply. Nevertheless, in this case greenhouse gas emissions and air pollution would increase sharply and impact on the environment could be considerable;

- According to the LGS scenario, GHG and dust emissions retain their levels, but consumers' demands for electricy would not be fulfilled, so that Kosovo's electric power system would not be a self-sufficient one;
- From analysis of these scenarios it follows that the most optimal and most realizable scenario is the medium growth scenario MGS, i.e. building of TPP New Kosova with capacity 600 MW and of capacities with renewable energy with at least 10% of the total electricity generation.

ACKNOWLEDGEMENTS

We would like to acknowledge the financial support of the Ministry of Education, Science and Technology of Republic of Kosovo.

REFERENCES

1. International Institute for Applied Systems Analysis, Energy Modeling Framework: Model for Energy Supply Strategy Alternatives and their General Environmental impact (MESSAGE).
2. Ea Energy Analyses, The STREAM modelling tool.
3. U.S. Energy Information Administration, Annual Energy Outlook with projections to 2040, Washington, 2013.
4. Mohammadi, H., Electricity Prices and Fuel Costs: Long-run Relations and Short-run Dynamics, *Energy Economics*, Vol. 31, No. 3, pp 503-509, 2009.

5. Krajačić, G., Duić, N., Carvalho, M. d. G., H$_2$RES, Energy Planning Tool for Island Energy Systems – The Case of the Island of Mljet, *Int. J. Hydrogen Energy*, Vol. 34, No. 16, pp 7015-26, 2009.
6. Connolly, D., Lund, H., Mathiesen, B. V., Leahy, M., A Review of Computer Tools for analysing the Integration of Renewable Energy into Various Energy Systems, *Applied Energy*, Vol. 87, No. 4, pp 1059-1082, 2010.

7. STELLA Software, High Performance Systems.
8. Forrester, J. W., *World Dynamics*, Cambridge, Massachusetts, Wright - Allen, Press Inc., 1971.
9. International Energy Agency, *Key World Energy Statistics*, IEA, Paris, 2012.
10. ExxonMobil, The Outlook for Energy: A View To 2040, Exxon Mobil Corporation, 2013.
11. International Energy Agency, CO$_2$ Emissions from Fuel Combustion, Highlights Version from International Energy Agency, Paris, 2012.
12. Forecast of Energy Demand in Kosovo for the Period 2007–2016, Ministry of Energy and Mining MEM, 2007.
13. Zidanšek, A., Blinc, R., Jeglič, A., Kabashi, S., Bekteshi, S. and Šlaus, I., Climate Changes, Biofuels and the Sustainable Future, *International Journal of Hydrogen Energy*, Vol. 34, Issue 16, pp 6980-6983, 2009.

14. Bekteshi, S., Kabashi, S., Šlaus, I., Zidanšek, A. and Najdovski, D., Modelling rapid climate changes and analyzing their impacts, *Management of Environmental Quality, An International Journal*, Vol. 19, pp 422-432, 2008.

15. Estimation of Population of Kosovo 2012, Kosovo Agency of Statistics, 2013.
16. The Energy Strategy of Republic of Kosovo 2009-2018, Ministry of Energy and Mining MEM, 2009.
17. Long Term Energy Balance of the Republic of Kosovo 2013-2022, Ministry of Economic Development, 2012.
18. Bekteshi, S., Kabashi, S., Podrimqaku, K., Šlaus, I. and Zidanšek, A., Scenario-based Analysis of Kosovo's Power Sector and CO₂ and Air Pollutants Emissions Reductions, *Proceeding of the Sustainable Development of Energy, Water and Environment Systems*, 8th Dubrovnik Conference, Dubrovnik, Croatia, 2013.

APPENDIX

Integrated electricity supply–demand and emission model as represented in STELLA model diagram

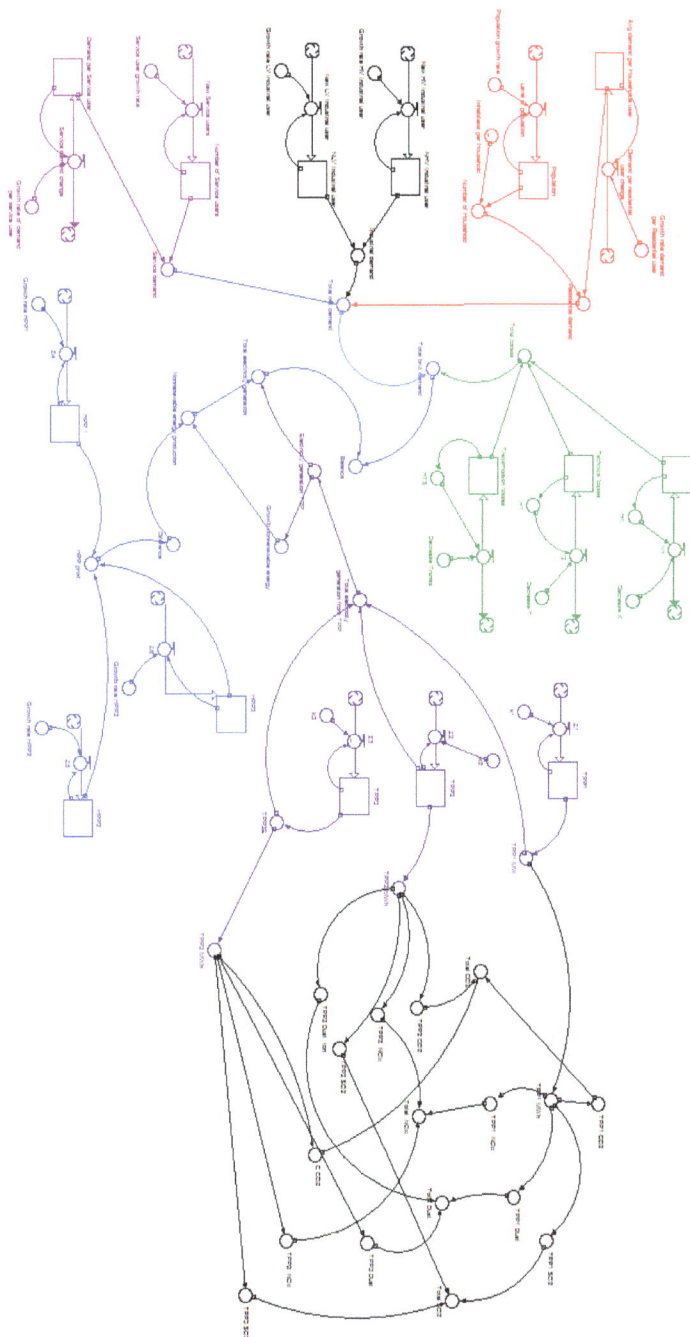

15

Testing Informational Efficiency in the EU ETS

Sebastian Goers
Energy Institute at the Johannes Kepler University, Linz, Austria
Department of Energy Economics
e-mail: goers@energieinstitut-linz.at

ABSTRACT

The paper deals with the analysis of informational efficiency of the European emissions trading scheme (EU ETS) with the goal of stating whether or not the system has been able to achieve its proclaimed cost-efficiency within the first two trading periods. The efficient market hypothesis suggests that profiting from predicting price behaviour is difficult as the market price should incorporate all available information at any time. I analyse the EU emission market to see if it shows evidence of the weak form of informational efficiency. In order to analyse the weak form of informational efficiency assessments I analyse random walk properties such as, the unit root, autocorrelation and variance ratio tests. The results reveal the existence of informational efficiency only in the second trading period.

KEYWORDS

European emissions trading scheme, Informational efficiency, Efficient market hypothesis, Random walk theory, Unit root tests, Autocorrelation, Variance ratio tests.

INTRODUCTION

The main objective of emissions trading is to offer a cost-efficient market-based instrument for emission mitigation. Cost-efficiency of the system induces that the predefined emission target is achieved by minimum costs and involves a well-functioning and established market, i.e. a market that provides informational efficiency. Informational efficiency implies that allowance prices display all significant information, market participants understand the realization of the emissions price and a forecast of future emissions prices, and hence earning above average returns is impossible. In the case of inefficient markets, due to transaction costs, incomplete information and heterogeneous expectations of market participants, which do not allow for forecasting future carbon price, an intensification of regulation in order to increase information flows and decrease market manipulation will be necessary. The fact that the European emissions trading scheme (EU ETS) finished the second trading phase at the end of 2012 allows for a complete investigation of the system's informational efficiency during the first two trading periods and to evaluate the scheme's effectiveness and efficiency during this timeframe.

The remainder of this paper is organized as follows. After some background information on the EU ETS and a brief review of econometric literature on the (informational) efficiency of the scheme in the introductory section, the second section explains theoretically the efficient market hypothesis and its connection to random walk theory. Thirdly, the statistical methodology and the data are described. The fourth section presents the results and the final section summarises the main findings.

Emissions trading in Europe

As an important player in the Kyoto and post-Kyoto process, the European Union (EU) itself decided to base its climate policy primarily on emission trading by the adoption of Directive 2003/87/EC establishing a scheme for greenhouse gas emission allowance trading within the Community. The latest amendment by Directive 2009/29/EC set the course for the time beyond the Kyoto Protocol period, originally based on the assumption that a global and comprehensive post-2012 agreement would be concluded in due time. The EU ETS is based on the cap-and-trade principle and offers tools to facilitate the achievement of climate targets without spoiling the economic scope of action. Firstly, by means of setting an emission cap, a price is put on carbon, which, secondly, enables trading GHG emission allowances on the carbon market at lowest cost.

Given the overall cap and using a downstream approach, the central EU authority, the EU Commission, has specified the trading sectors of the economy where emission allowances are traded: iron and steel, cement, glass, ceramics, pulp and paper, as well as the power sector. These sectors account for around 50% of the EU emissions of carbon dioxide (CO_2) and for 40% of the EU's overall greenhouse gas emissions. Because 50% of the CO_2 EU emissions remain outside the trading program, the EU's Kyoto cap has to be met by an effort-sharing arrangement between sources in the trading and non-trading sectors. Industries that are not covered by the scheme, such as the private sector, transport, or the building industry, have to be regulated by other (national) abatement measures in order to reach each national emission reduction target. The EU ETS established three commitment periods for the time up to 2020. The first trading period (2005-2007) was actually seen as a test run and a learning phase to find out how the different actors involved in emissions trading react to the new system. The second phase (2008-2012) corresponded to the Kyoto commitment period. Within these two trading periods, the Member States endowed domestic covered sectors via the so-called National Allocation Plans (NAPs) with emissions allowances, which were subject to oversight from the European Commission. As far as the ETS sectors are concerned, the EU has learned important lessons for its third trading period (2013-2020) regarding the system's economic efficiency and ecological effectiveness. Hence, cap settings as well as harmonised allocation rules are determined directly at the EU level pursuant to the revised ETS directive. Further, within the revision and the preparation for a post-Kyoto period of the EU ETS, one central point was the intensification of auctioning allowances from 2013.

Economic literature on the efficiency of the EU ETS

In general, econometric literature on carbon pricing via emissions trading and especially the prices generated through the EU ETS is growing. Analyses show that daily spot prices generated by the EU ETS depend on institutional design issues, energy prices and extreme weather events [1]. The approach by Chevallier explains that CO_2 future prices of the EU ETS are only weakly connected to macroeconomic effects [2]. In a different model of Creti et al. the oil price, the equity price index and the switching price between gas and coal seem to be significant long-run determinants of the EU CO_2 future price during the second phase of the EU ETS [3]. Further, the decisions of the second NAPs have crucial and direct influences on the EU ETS future prices [4].

Regarding the analysis of informational efficiency of the EU ETS until 2010, Aatola et al. find that the EU ETS market showed periods with no informational efficiency [5, 6]. It is shown by Daskalakis and Markellos that three of the most important spot and future markets for EU ETS CO_2 allowances deviate from the weak form of market efficiency [7]. Joyeux and Milunovich [8] do not detect informational efficiency in the EU ETS market in the first trading period while Hintermann [9] discovers inefficiency in the EU ETS before the price crash in 2006. In contrast to that, phases of informational efficiency within the first trading period have been noted. Seifert *et al.* use a stochastic optimal control model and show that CO_2 prices do not follow seasonal patterns and that the EU ETS market worked informationally efficient [10]. Boutaba presents evidence that the EU ETS, among different carbon markets, showed a reasonable degree of efficiency in the short and long term [11]. Regarding a wider timeframe until spring 2009, Montagnoli and de Vries [12] observe the weak form of informational efficiency in the second phase of the EU ETS. Krishnamurti and Hoque [13] identify informational efficiency in the EU ETS CO_2 option prices and that short maturity options are priced more efficiently than distant maturity options in the first half of the second trading period. In contrast to the stated literature, the present approach covers the whole of the first two trading periods, giving a wider base to the analysis which allows for a more accurate assement of informational efficiency in the EU ETS.

EFFICIENT MARKET HYPOTHESIS AND RANDOM WALK THEORY

The efficient market hypothesis (EHM), also known as the concept of informational efficiency or capital market efficiency derived by Fama, is an economic theory which deals with the information processing in capital markets [14]. A market is efficient if "security prices at any time "fully reflect" all available information" which means that all information is already factored in the actual price and hence, the realization of excess returns via forecasting methods is not possible [14]. This implies that in an informationally-efficient market not just present and historical data but also anticipated developments are taken into account within the price formation process. Shifting the focus to the EU ETS, this implies that market participants are aware of the relevant CO_2 price data and the CO_2 price data generating processes. In the case of new information, the market participants re-evaluate the shareholder values which lead to new CO_2 price levels. As information is only classified to be new in case of non-anticipation of the market participant, shocks which directly affect the investment behaviour or the data generating processes can influence the price level.

Three diverse categories of information which yield three forms of the EHM can be identified [14]. Strong informational efficiency implies that the CO_2 price reflects all available information. This assumption yields that publicly available and non-publicly available information (e.g. executive board development, mergers) is factored into the price analysis. The category of semi-strong informational efficiency advocates that actual CO_2 price levels fully reflect information which is publicly available. Hence, all historical and fundamental data (e.g. economy, weather, fossil fuel prices) is integrated into the price signal. Therefore, only the usage of non-publicly available information or inside information allows for the generation of above-average returns. Event studies measuring the velocity of price changes due to new information can be applied to test this form of market efficiency. Finally, the weak form of the EHM proclaims that the actual price fully includes information of historical prices and returns which do not have any influence in future price developments. This suggests that regarding the weak informational efficiency of the EU ETS an analysis of past CO_2 price behaviour as done by technical analysis does not lead to the generation of above-average returns. Only the availability of additional information allows generating higher returns. This form of efficiency is tested via analysis of the predictability of future returns with historical price data, as is done in this paper. The test of the weak form of informational efficiency implies that impacts of other variables than historic price levels on the CO_2 price are disregarded. If effects of other variables are not directly reflected in the actual price level but rather influence the price level step by step, market participants will be informed about smaller rises about to come. Therefore, the reason for the change of the CO_2 price change is unnecessary within this framework as the question is if and to what extend past price changes are informative for future price changes.

The weak-form of the EHM is linked to the statistical concept of a random walk, which states that all subsequent price changes symbolize random deviations from earlier prices. The random walk hypothesis assumes that the flow of information is unrestricted and information is directly integrated into market prices meaning that future prices will be independent of present price changes. Newly arriving information cannot be predicted which leads to the consequence that price changes are unpredictable and random. A random walk is defined as a stochastic process in the form of an autoregressive process, $p_t = p_{t-1} + \beta + \varepsilon_t$, where p_t symbolizes the natural logarithm of the EUA price at point of time t, β denotes a drift parameter and ε_t represents the random increment and is independent and identically distributed with mean zero and variance σ^2. The random term can be interpreted as the effect of arriving information on the actual CO_2 price. The first difference is displayed via $\Delta p_t = \beta + \varepsilon_t$. Further, this model states that the expected value of the CO_2 price is identical to the expected value in the previous periods adjusted to the unanticipated information.

ANALYSIS OF INFORMATIONAL EFFCICENCY OF THE EU ETS

In the following, the underlying statistical methodology for random walk testing and the examined data set are presented.

Statistical methodology

According to former approaches [6, 15] the focus is laid on analysing random walks which are characterised by dependent and not identically distributed random increments. Thus, the empirical methodology to investigate the informational efficiency contains unit root tests, analysis of autocorrelation and variance ratio tests.

Unit root tests. Unit root tests are used to examine whether a time series represents a non-stationary stochastic process. As a random walk is a first difference stationary process, CO_2 prices generated by the EU ETS need to contain a unit root while the first difference of the series does not. The Augmented Dickey-Fuller Test (ADF) is an augmented form of the Dickey-Fuller Test (DF) [16] which responds to larger and more complex time series models. It extends the framework of the DF in the sense of assuming that p_t follows an autoregressive process of order k with c denoting a constant and $k > 1$:

$$p_t = c + \delta_t p_{t-1} + ... + \delta_k p_{t-k} + \varepsilon_t \qquad (1)$$

This process is equal to:

$$\Delta p_t = c + \theta p_{t-1} + \beta_1 \Delta p_{t-1} + ... + \beta_{k-1} \Delta p_{t-(k-1)} + \varepsilon_t \qquad (2)$$

where $\theta := \delta_1 + ... + \delta_{k-1}$. $\theta < 0$ holds for a stationary process while $\theta = 0 \Leftrightarrow \delta := \sum \delta_i = 1$ indicates a non-stationary process. Hence, the null hypothesis which says that p_t is non-stationary and contains a unit root is given by:

$$H_0 : \theta = 0 \qquad (3)$$

The alternative hypothesis stating that p_t is a stationary process is represented by:

$$H_1 : \theta < 0 \tag{4}$$

The ADF requires homoskedastic and independent error terms ε_t. Regarding the choice of the lag-length of the autoregressive model, it should be chosen sufficiently large in order to avoid misspecification of the model. On the contrary, a too generous k may lead to the fact that the null hypothesis is not rejected. In practice, k should be set in the way that a defined criterion of information which allows comparing the validity of nested models is fulfilled. In this case, the Akaike-Information criterion (AIC) is applied.

A further approach to test for unit roots is given by the Phillips-Perron Test (PP) [17]. The PP adjusts for serial correlation and heteroskedasticity in the errors terms non-parametrically by adapting the DF test statistics. Therefore, these test statistics can be interpreted as DF test statistics adaptive to serial correlation by applying the Newey-West heteroskedasticity - and autocorrelation-consistent covariance matrix estimator. Besides the advantage of allowing for heteroskedasticity in the error term, the further benefit is that no lag length for the test regression has to be specified compared to the ADF.

Finally, the methodology of the Kwiatkowski-Phillips-Schmidt-Shin Test (KPSS) [18] is also applied within the testing for unit roots. In contrast to the procedures of the ADF and the PP, the KPSS proposes stationarity as the null hypothesis and non-stationarity as the alternative hypothesis. The time series is modelled as the sum of a deterministic trend, a random walk and a white noise whereas the KPSS tests if the random walk has a zero variation using specific critical values.

Autocorrelation coefficients. In order to state if the CO_2 price series follows as a random walk the next step contains an autocorrelation analysis. Autocorrelation of a series regarding the k-th lag refers to the correlation between the lags p_t and p_{t-k}. Regarding the random walk model with dependent and not identically distributed random increments, all autocorrelation coefficients between Δp_t and Δp_{t-k} need to equal zero for all $k>0$. Hence, the k-th autocorrelation coefficient, $\rho(k)$, can be described as:

$$\rho(k) = \frac{\text{cov}(p_t - p_{t-k})}{\sqrt{\text{var}(p_t)\text{var}(p_{t-k})}} \tag{5}$$

In the case of serial autocorrelation, an autoregressive model of order j displays p_t as a linear function of the lagged variables $p_{t-1},...,p_{t-j}$. Autocorrelation coefficients going to zero continuously with a growing k indicate autoregressive properties of the process whereas an abrupt reduction of the coefficients to zero in case of a growing k represents a moving-average process. The application of the Ljung-Box Q-statistic at lag k allows for testing the null hypothesis that there is no autocorrelation up to order k and the significance of the autocorrelation coefficients, respectively.

Variance ratio tests. As stated in the previous subsection, the autocorrelation $\rho(k)$ between Δp_t and Δp_{t-k} needs to equal zero for all $k>0$ in order to fulfil the requirements of a random walk with dependent and not identically distributed random increments. The variance ratio test indicates if the variance of the random walk's error term is a linear function of time. The variance ratio derived by Lo and MacKinlay [19] is defined as:

$$VR(q) = \frac{\text{var}(\Delta p_t(q))}{q\text{var}(\Delta p_t)} = 1 + 2\sum_{k=1}^{q-1}(1-\frac{k}{q})\rho(k) \tag{6}$$

The standardized test statistic ψ investigates the independence of the error terms of the process. It is defined as:

$$\psi = \frac{\sqrt{T-1}(VR(q)-1)}{\sqrt{\hat{\theta}}} \sim N(0,1) \tag{7}$$

θ represents a heteroscedasticity-consistent estimator of the variance of $VR(q)$.

Sample structure and data transformation

In order to study the behaviour of the EUA prices, daily settlement price data from August 2005 to June 2012 is analysed which leads to 1,468 observations. The data is based on price information provided by the European Energy Exchange (EEX). Structuring the data with regard to the first and second trading phases, the sample of the first trading phase includes 607 observations from August 2005 to December 2007 while the sample of the second trading phase includes 861 observations from January 2009 to June 2012. Due to missing data on the settlement prices from March 2008 to December 2008, the sample of the EUA price for the second trading phase disregards data for 2008. Figure 1 displays the price levels in the different trading phases.

As already mentioned in the introductory section, overallocation due to strategic allowance endowment to domestic EU ETS sectors by governments led to the abrupt price drop in spring 2006 [1]. Until April 2007, the EUA price almost arrived at €0.1 per ton CO_2. In the second trading period (2008-2012), less volatile price behaviour is observed. Stricter revision of the NAP and, consequently, more stringent cap setting by the EU Commission led to an average spot price of €14 per ton CO_2 in 2009 and 2010, €13 per ton CO_2 in 2011 and €7 per ton CO_2 in 2012. Regarding this second trading phase, economic recession inducing lower production activities by the covered sectors and unsuccessful climate policy negotiations may have generated low EUA price levels.

In the present analysis, EUA prices at point of time t are denoted by P_t and are examined via its natural logarithm series, $p_t = \ln(P_t)$, and the differentiated natural logarithm series, $\Delta p_t = p_t - p_{t-1}$, displaying proportional deviations of the original price series. Hence, Δp_t expresses the logarithmic EUA price returns at point t. Figures 2 and 3 show the price history of p_t and Δp_t while the series' descriptive statistics are presented in Table 1.

Table 1. Descriptive statistics for p_t and Δp_t

	p_t		Δp_t	
Trading period	2005-2007	2009-2012	2005-2007	2009-2012
Mean	1.408	2.508	-0.011	0.007
Variance	5.735	0.070	0.010	0.049
Kurtosis	1.881	3.270	34.245	833,832
Skewness	-0.714	1.200	1.346	28.644
Observations	607	861	606	860

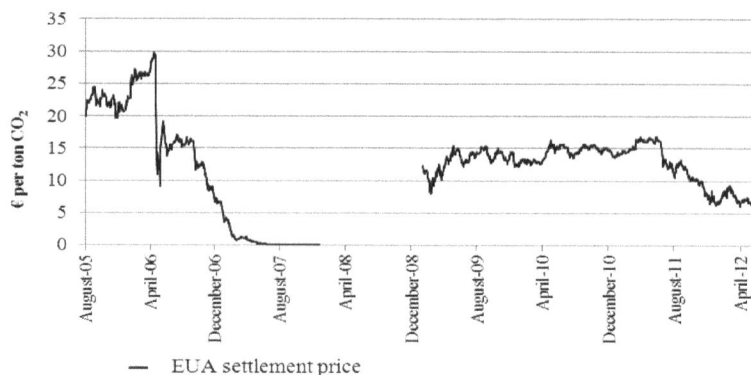

Figure 1. EUA settlement prices in the EU ETS

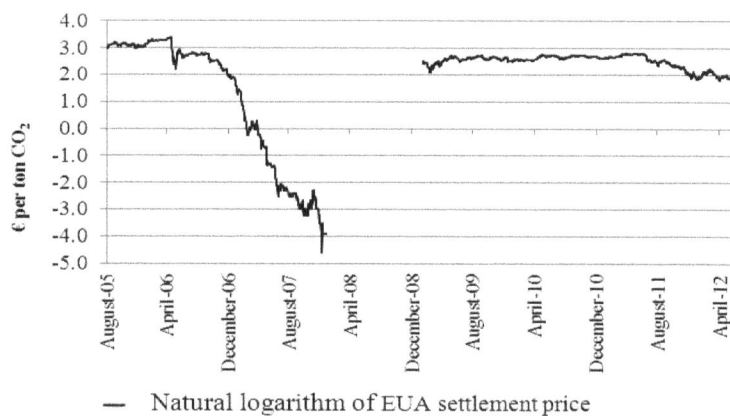

Figure 2. Natural logarithm of EUA settlement prices in the EU ETS

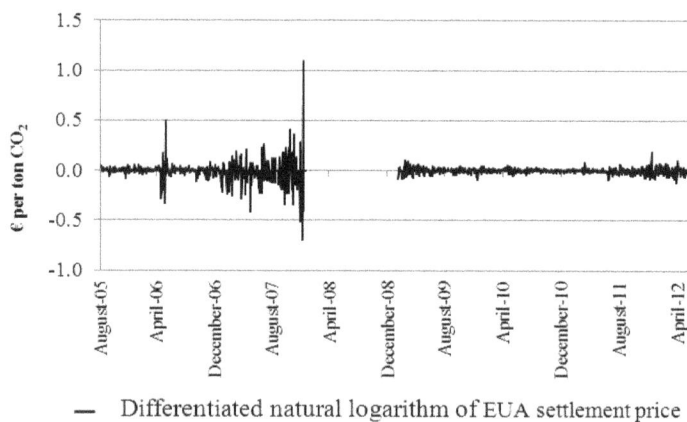

Figure 3. Differentiated natural logarithm of EUA settlement prices in the EU ETS

RESULTS

Based on the described statistical methodology this section presents the results and offers an interpretation regarding the scheme's level of informational efficiency in the focused timeframes. As explained above, unit root tests are used to examine whether a time series represents a non-stationary stochastic process. The presence of a unit root is interpreted as the null hypothesis. As random walks are first difference stationary processes, if p_t contains a unit root it is non-stationary, while if Δp_t does not contain a unit root it is stationary. As Table 2 reports, the performed unit root tests indicate at 10%, 5% and 1% significance levels non-stationarity for p_t and stationarity for Δp_t regarding the first trading period (2005-2007). Considering the second trading phase (2009-2012), only the ADF-test ignoring a constant and a trend does not reject the null of non-stationarity for p_t at a 10% significance level. On the contrary, the KPSS-tests reject clearly stationarity for p_t at all significance levels. Analogously to the first trading phase, the unit root tests reveal stationarity for Δp_t at all significance levels. Hence, the results point out that CO_2 prices generated by the EU ETS may have followed a random walk during the first two trading periods whereas the findings for the first trading period show stronger evidence.

Table 2. Unit root tests for p_t and Δp_t

Test	Test statistics for p_t		Test statistics Δp_t	
	2005-2007	2009-2012	2005-2007	2009-2012
ADF (constant)	1.49	-23.55^{***}	-4.00^{***}	-10.40^{***}
ADF (constant+trend)	-1.64	-30.89^{***}	-4.66^{***}	-10.60^{***}
ADF	-0,20	-1.60^{*}	-2.88^{***}	-10.36^{***}
PP (constant)	1.61	-27.67^{***}	-30.03^{***}	-29.70^{***}
PP (constant+trend)	-1.66	-30.01^{***}	-30.16^{***}	-29.80^{***}
PP	0.12	-1.64^{***}	-29.14^{***}	-29.67^{***}
KPSS (constant)	2.96^{***}	1.47^{***}	0.65	0.38
KPSS (constant+trend)	0.76^{***}	0.72^{***}	0.07	0.12

*, ** and *** refer to significance at the 10%, 5% and 1% levels.
Regarding the ADF-tests, the number of lags is specified via the AIC.

In order to adopt the random walk hypothesis, autocorrelation between Δp_t and Δp_{t-k} for all $k > 0$ needs to equal zero. The results of the autocorrelation analysis as well as their significance indicated by the Ljung-Box $Q(k)$-statistic are presented in Table 3. With respect to the autocorrelation of p_t in the first and the second trading period, strong positive autocorrelation can be observed by gradually decreasing coefficients with increasing number of lags. Studying the autocorrelation of Δp_t within the first trading period (2005-2007) leads to the conclusion that the null hypothesis of no autocorrelation has to be rejected at all significance levels. Regarding the second trading period (2009-2012), autocorrelation coefficients near to zero and low Ljung-Box $Q(k)$-statistics reveal that lagged change in the logarithmic CO_2 price does not explain the current change. As a conclusion, the requirements that the EUA price followed a random walk are only fulfilled in the second trading period (2009-2012).

Finally, variance ratio tests are applied in order to study the EUA price series' incremental behaviours. Following the random walk hypothesis, the increments need to follow a linear function of time. This means that the q-period difference should be q times the variance of the one-period difference. The results in Table 4 display that random walk properties are not satisfied in the first trading period (2005-2007). The null of a random walk can be rejected for all $q \leq 7$ at all significance levels. Nevertheless, for $q > 10$ the test fails to reject the null of no significant autocorrelation among the returns. Focusing the second trading period (2009-2012), values of $VR(q)$ are near to one until $q = 10$ and throughout we fail to reject the null of autocorrelation, indicating the presence of a random walk. Hence, the variance ratio tests support the findings of the former autocorrelation analysis which found random walk behaviour in CO_2 prices for the second trading period.

Table 3. Autocorrelations for p_t and Δp_t

	p_t in 2005-2007		Δp_t in 2005-2007	
Lag k	Autocorrelation	$Q(k)$	Autocorrelation	$Q(k)$
1	0.995	604.0[***]	-0.194	22.87[***]
2	0.990	1,203.2[***]	0.010	22.93[***]
3	0.986	1,797.8[***]	0.035	23.69[***]
4	0.981	2,387.5[***]	-0.084	28.10[***]
5	0.976	2,972.5[***]	0.085	32.44[***]
6	0.971	3,552.6[***]	0.054	34.23[***]
7	0.967	4,128.4[***]	0.084	38.52[***]
8	0.961	4,697.9[***]	-0.119	47.32[***]
9	0.956	5,262.4[***]	0.098	53.30[***]
10	0.950	5,821.7[***]	-0.114	61.28[***]
11	0.945	6,376.0[***]	0.028	61.78[***]
12	0.941	6,926.0[***]	0.093	67.17[***]
13	0.936	7,471.4[***]	-0.055	69.04[***]
14	0.932	8,012.4[***]	-0.054	70.29[***]
15	0.927	8,549.0[***]	0.048	71.74[***]
16	0.923	9,081.5[***]	-0.060	73.98[***]
17	0.918	9,609.7[***]	-0.011	74.06[***]
18	0.914	10,134[***]	0.088	78.87[***]
19	0.910	10,655[***]	-0.025	79.26[***]
20	0.906	11,172[***]	-0.061	81.59[***]
	p_t in 2009-2012		Δp_t in 2009-2012	
Lag k	Autocorrelation	$Q(k)$	Autocorrelation	$Q(k)$
1	0.992	849.54[***]	-0.012	0.12
2	0.982	1,684.5[***]	-0.005	0.14
3	0.974	2,506.0[***]	0.001	0.14
4	0.965	3,313.9[***]	0.005	0.16
5	0.957	4,108.1[***]	-0.002	0.17
6	0.948	4,889.3[***]	0.004	0.18
7	0.940	5,658.7[***]	-0.000	0.18
8	0.932	6,414.9[***]	-0.000	0.18
9	0.923	7,157.7[***]	0.001	0.18
10	0.915	7,888.1[***]	-0.001	0.18
11	0.907	8,607.2[***]	-0.005	0.20
12	0.900	9,315.3[***]	-0.013	0.35
13	0.893	10,014[***]	-0.002	0.35
14	0.885	10,701[***]	-0.003	0.36
15	0.877	11,377[***]	-0.002	0.36
16	0.869	12,041[***]	-0.000	0.36
17	0.861	12,694[***]	-0.001	0.46
18	0.852	13,333[***]	-0.012	0.58
19	0.844	13,962[***]	-0.009	0.66
20	0.837	14,580[***]	0.007	0.71

*, ** and *** refer to significance at the 10%, 5% and 1% levels.

Table 4. Variance ratio test for Δp_t

	2005-2007		2009-2012	
q	$VR(q)$	$\Psi(q)$	$VR(q)$	$\Psi(q)$
2	0.860	-4.777***	0.987	-0.374
3	0.747	-4.171***	0.979	-0.415
4	0.735	-3.481***	0.975	-0.394
5	0.695	-3.432***	0.974	-0.351
6	0.695	-3.034***	0.972	-0.332
7	0.709	-2.632***	0.955	-0.490
8	0.730	-2.250**	0.984	-0.159
9	0.730	-2.094**	0.981	-0.172
10	0.750	-1.824*	0.980	-0.178
20	0.751	-1.232	0.863	-0.809

*, ** and *** refer to significance at the 10%, 5% and 1% levels.

CONCLUSION

The concept of informational efficiency proclaims that the market price should include all available information at any time. Firms subject to the EU ETS face uncertainty regarding their investment and production activities which suggests that competitive disadvantages may occur in relation to firms which are not regulated or face credible CO_2 signals. Hence, the inexistence of informational efficiency constrains the object of cost-efficiency which requires that emissions abatement is achieved at lowest costs. The focus is drawn on the weak form of informational efficiency which suggests that price fully includes information of historical prices and returns which do not have any influence in future price developments.

By applying unit root, autocorrelation and variance ratio tests, evidence is derived that within the first trading period the EU ETS did not operate with informational efficiency whereas it did in the second trading period. This implicates that within the second trading period market participants had a better understanding of the CO_2 price generating processes and the way in which information affects the equilibrium CO_2 price. Compared to existing literature on the informational efficiency of the EU ETS, these findings support past findings of no informational inefficiency within the first trading period of Dsakalakis and Markellos [7], Hintermann [9], Joyeux and Milunovich [8], Montagnoli and de Vries [12]. Further, as the analysis covers the complete second trading period, the findings add the first evidence of informational efficiency in first parts of the second trading period achieved by Montagnoli and de Vries [12]. Further research from a financial economics point of view regarding the scheme's informational efficiency within the third trading period will be necessary to state if the EU ETS is on a continuous path to better efficiency via the regulatory changes of 2013 which included, centralised cap setting and a harmonisation of allocation rules.

ACKNOWLEDGEMENTS

I would like to thank the editor, three anonymous reviewers and Friedrich Schneider for their valuable comments on previous versions of this article. Further, the support of this work by the Energy Institute at the Johannes Kepler University of Linz is gratefully acknowledged.

NOMENCLATURE

c	constant	[-]
k	order of process, lag	[-]
P	European CO_2 price	[-]
p	natural logarithm of European CO_2 price	[-]
q	period	[-]
VR	variance ratio	[-]

Greek letter

β	drift parameter	[-]
Δ	first difference	[-]
ε	error term	[-]
σ	variance	[-]
ρ	autocorrelation coefficient	[-]
Ψ	standardized test statistic	[-]

Subscript

t	point of time	[-]

Abbreviations

ADF	Augmented Dickey-Fuller Test
AIC	Akaike-Information Criterion
CO_2	Carbon Dioxide
DF	Dickey-Fuller Test
EEX	European Energy Exchange
EHM	Efficient Market Hypothesis
EU	European Union
EUA	European Allowance
EU ETS	European Emissions Trading Scheme
KPSS	Kwiatkowski-Phillips-Schmidt-Shin Test
NAP	National Allocation Plan
PP	Phillips-Perron Test

REFERENCES

1. Alberola, E., Chevallier, J. and Cheze, B., Price Drivers and Structural breaks in European Carbon Prices 2005-2007, *Energy Policy*, Vol. 36, No. 2, pp 787-79, 2008.

2. Chevallier, J., Carbon Futures and Macroeconomic Risk Factors: A view from the EU ETS, *Energy Economics*, Vol. 31, No. 4, pp 614-625, 2009.

3. Creti, A., Jouvet, P. and Mignon, V., Carbon Price Drivers: Phase I versus Phase II Equilibrium? *Energy Economics*, Vol. 34, No. 1, pp 327-334, 2012.

4. Conrad, C., Rittler, D. and Rotfuss, W., Modeling and Explaining the Dynamics of European Union Allowance Prices at High-frequency, *Energy Economics*, Vol. 34, No. 1, pp 316-326, 2012.

5. Aatola, P., Ollikka, K. and Ollikainen, M., Weak and Semi-Strong Forms of Informational Efficiency in the EU ETS Markets, *Discussion Papers no 48*, University of Helsinki, Department of Economics and Management, Helsinki, 2010.

6. Aatola, P., Ollikka, K. and Ollikainen, M., Informational Efficiency of the EU ETS Market-a Study of Price predictability and Profitable trading, *Working Papers no 28,* Government Institute for Economic Research, Helsinki, 2012.

7. Daskalakis, G. and Markellos, R., Are the European Carbon Markets Efficient? *Review of Futures Markets*, Vol. 17, No. 2, pp 103-128, 2008.

8. Joyeux, R. and Milunovich, G., Market Efficiency and Price Discovery in the EU Carbon Futures Market, *Applied Financial Economics*, Vol. 20, No. 10, pp 803-809, 2010.

9. Hintermann, B., Allowance Price drivers in the First Phase of the EU ETS, *Journal of Environmental Economics and Management*, Vol. 59, No. 1, pp 43-56, 2010.

10. Seifert, J., Uhrig-Homburg, M. and Wagner, M., Dynamic Behavior of CO_2 Spot Prices, *Journal of Environmental Economics and Management*, Vol. 56, No. 2, pp 180-194, 2008.

11. Boutaba, M., Dynamic Linkages among European Carbon Markets, *Economics Bulletin*, Vol. 29, No. 2, pp 499-511, 2008.

12. Montagnoli, A. and de Vries, F., Carbon Trading Thickness and Market Efficiency, *Energy Economics,* Vol. 32, No. 6, pp 1331-1336, 2010.

13. Krishnamurti, C. and Hoque, A., Efficiency of European Emissions Markets: Lessons and Implications, *Energy Policy*, Vol. 39, No. 10, pp 6575-6582, 2011.

14. Fama, E., Efficient Capital Markets: A review of Theory and Empirical Work, *Journal of Finance,* Vol. 25, No. 2, pp 383-417, 1970.

15. Albrecht, J., Verbeke, T. and De Clerq, M., Informational Efficiency of the US SO_2 Market, *Environmental Modeling and Software*, Vol. 21, No. 10, pp 1471-1478, 2006.

16. Dickey, D. and Fuller, W. Distribution of the Estimators for Autoregressive Time Series With a Unit Root, *Journal of the American Statistical Association*, Vol. 74, pp 427-431, 1976.

17. Philips, P. and Perron, P., Testing for a Unit Root in Time Series Regression, *Biometrika*, Vol. 75, No. 2, pp 335-346, 1988.

18. Kwiatkowski, D., Phillips, P., Schmidt, P. and Shin, Y., Testing the Null Hypothesis of Stationarity against the Alternative of a Unit Root, *Journal of Econometrics*, Vol. 54, No. 1-3, pp 159-178, 1992.

19. Lo, A. and MacKinlay, A., Stock Market Prices do not Follow Random walks: Evidence from a Simple Specification Test, *The Review of Financial Studies*, Vol. 1, No. 1, 41-66, 1988.

16

Wuxi – a Chinese City on its Way to a Low Carbon Future

*Carmen Dienst[*1], Chun Xia[1], Clemens Schneider[1], Daniel Vallentin[1], Johannes Venjakob[1], Ren Hongyan[2]*

[1]Wuppertal Institute for Climate, Environment and Energy GmbH, Wuppertal, Germany
e-mail: carmen.dienst@wupperinst.org
[2]Jiangnan University, Environment and Civil engineering school, Wuxi Low Carbon Development Research Centre (WLCDC), Wuxi, China

ABSTRACT

Urbanization and climate change are amongst the greatest challenges of the 21st century. In the "Low Carbon Future Cities" project (LCFC), three important problem dimensions are analysed: current and future GHG emissions and their mitigation (up to 2050), resource use and material flows and vulnerability to climate change. The industrial city of Wuxi has been the Chinese pilot city of the project. To establish the pathway for a low carbon future, it is crucial to understand the current situation and possible future developments. The paper presents the key results of the status quo analysis and the future scenario analysis carried out for Wuxi. Two scenarios are outlined. The Current Policy Scenario (CPS) shows the current most likely development in the area of energy demand and GHG emissions until 2050. Whereas the extra low carbon scenario (ELCS) assumes a significantly more ambitious implementation, it combines a market introduction of best available technologies with substantial behavioural change. All scenarios are composed of sub-scenarios for the selected key sectors. Looking at the per capita emissions in Wuxi, the current levels are already high at around 12 tonnes CO_2 per capita compared to Western European cities. Although Wuxi has developed a low carbon plan, the projected results under current policies (CPS) show that the total emissions would increase to 23.6 tonnes CO_2 per capita by 2050. If the ELCS pathway was to be adopted, these CO_2 emission levels could be reduced to 6.4 tonnes per capita by 2050. However, this is not a problem unique to Wuxi or China. A comprehensive rethink at global level on how to increase energy and resource efficiency and sustainability is required.

KEYWORDS

Low carbon future cities, GHG inventory, Low carbon city strategies, China, Wuxi, Low carbon scenario, Low carbon technologies, Mitigation.

INTRODUCTION

Urbanization and climate change are amongst the greatest challenges of the 21st century and urban areas are key sources of global GHG emissions, with an estimated share of 40 to 78% [1]. Cities are also very likely to face the impacts of climate change and need to learn to adapt. In addition, scarce natural resources further constrain the scope for long-term, sustainable urban development.

The LCFC project

The Low Carbon Future Cities (LCFC) project aims to overcome different challenges in three dimensions - besides GHG emissions and their mitigation, resource use and

---

[*] Corresponding author

vulnerability to climate change are also analysed. For the pilot city Wuxi (China) project, the research project team has developed an integrated strategy and roadmap for balancing low carbon development, improved resource efficiency and adaptation to climate change [2].

Initially, the project focused on an analysis of the status quo and current policies, as well as extrapolating the most likely resultant development pathway up to 2050 with regard to GHG emissions, energy demand, climate impacts and resource use [3, 4].

During this initial phase it became apparent that in order to effectively tackle all the problem dimensions in the future, further efforts would be required such as the application of a suitable advanced technology portfolio. Moreover, developing a low carbon future strategy for Wuxi calls for integrated tools that can address considerable complexities and uncertainties that are inherent in urban systems and their development. These considerations go beyond the short and medium-term horizon of common urban planning practices.

Therefore, two low carbon scenarios were developed. These scenarios applied the same technologies and policy measures, but with different levels of mitigation ambition, in an attempt to incorporate an understanding of long-term challenges and low carbon solutions [5, 6].

These scenario analyses help to enable policy makers to identify strategic areas for action and provide a basis for decision-making. In contrast to the CPS which reflects the current policies, these two alternative scenarios (LCTS and ELCS) reveal the need for action to achieve certain desirable targets and identify windows of opportunity. Thereby it is crucial to ensure that the scenarios assumptions are adapted to local conditions and the results can effectively contribute to the development of the integrated low carbon strategy. The paper presents the key results of the status quo analysis and the future scenario analysis for the city of Wuxi.

Background to Wuxi

Wuxi is an industrial city with a population of about 6 million, located on the shores of Lake Taihu in the east of China near the megacity of Shanghai. As is the case of many Chinese cities, Wuxi's economy has increased considerably in recent years. The overall GDP in 2010 was about EUR 64.3 billion, which was double that of 2005. This represented an average annual growth rate of 14%. However, this growth has gone hand in hand with a significant increase in negative environmental effects, such as air pollution and GHG emissions. Wuxi is also exposed to climate and weather-related risks, such as extreme temperatures or floods. The city government was already quite aware of these challenges and had developed a low carbon plan. This made Wuxi a good choice for the Chinese pilot city project. The lessons learned from the project study can be used to analyse the situations in other similar Chinese industrial cities. Especially as former studies or low carbon initiatives focus on China's megacities.

METHODOLOGY

The analysis of the status quo and trend assessment, as well as the scenario development for Wuxi, was divided into different working areas, each applying a different type of methodology:

- Emission inventory: status quo and quantification of CO_2 emissions and qualitative analysis of non-CO_2 emissions (Kyoto gases) where possible;
- Key sector identification: mainly based on inventory, future trends and current policies;

- Scenario analysis: including Current Policy Scenario (CPS), Low Carbon Technology Scenario (LCTS) and Extra Low Carbon Scenario (ELCS).

Following a brief description of the methodology used in the GHG inventory and key sector selection, details are given on the scenario analyses and the assumptions applied.

GHG inventory

Over the last decades, much emphasis has been put on the development and implementation of GHG monitoring. However, these methods focused primarily on national inventories, such as in the case of the IPCC guidelines that were applied in this study in a simplified form. Although there is a lot of relevant literature in research, as well as practical handbooks/manuals at city level, there is still lack of standardized and globally applied methods and lack of reliable comparisons of GHG inventories between cities [7].

In China, several cities that are aiming to develop low carbon pathways have applied methods for GHG monitoring and quantified their impacts [7-9], but the comparison of the inventory results is only possible to a very limited extent. This is also due, in part, to the differing scope of emissions (territorial or supply chain and consumption-based approaches) and administrative boundaries considered in the applied methods, which can significantly affect the emission accounting [7]. In contrast to city boundaries in many other countries, the situation in China is different as the administrative city boundaries include not only the city but also non built-up areas such as the surrounding agricultural and forestry land [9].

The GHG emissions in the city of Wuxi are calculated and analysed in a detailed GHG inventory. The inventory has three main objectives:

- To provide a basis for key sector identification;
- To provide a basis for measuring future GHG reductions;
- To identify data gaps and options for developing a comprehensive and regularly updated inventory in the future.

The methodology for national inventories, outlined in the 2006 IPCC guidelines on GHG inventories [10], was followed in a simplified manner. The study focused mainly on CO_2 emissions. Other GHGs were included when data was available. The reporting year for Wuxi's emissions is 2009. For the energy sector, emissions from 2003 to 2009 were calculated.

Four main sectors were considered in accordance with the IPCC guidelines:

- Energy (including all energy-related emissions, also from industry);
- Industrial Processes and Product Use (IPPU) (only process-related emissions);
- Agriculture, Forestry and Other Land Use (AFOLU);
- Waste.

The study focused on CO_2 emissions; other GHGs were considered to a lesser extent depending on available data. According to the IPCC, the so-called "territory principle" was followed, meaning that GHG emissions were allocated to the territory where they were emitted. Life-cycle emissions and material use by specific sectors were not analysed in the inventory, but were included in the work on resource and material use [3, 5].

The inventory is largely based on Wuxi's statistical yearbook [8] and data provided by Chinese project partners, especially by the regional partner institution Wuxi Low Carbon Development and Research Centre (WLCC).

Although the statistical yearbook [11] provides extensive data which is exceptional at city level, there was still a gap between the data available and the data required for the inventory and the projection models. Particularly for the industry sector only statistical data for "above designed size" industries were available, which refers to those with an annual prime operating revenue exceeding CNY 5 million.

In addition to the incompleteness of available data, the available statistical data should be approached with caution. A study that was undertaken by a group of scientists from China, Britain and the United States analysed different sets of Chinese statistical data. The comparison showed that there are significant variances between the data and that these variances could account for a difference of up to 1.4 billion tonnes of CO_2 emissions (for 2010) between the datasets for the same area [12, 13], which is a difference of 18% in estimates of CO_2 emissions from China when using these two different data sets.

Modelling and scenario analysis

Scenario analysis, which has been widely applied in sustainability research over the last decade, is a tool that was used to explore different pathways for future developments in Wuxi and the associated energy demand and CO_2 emissions.

In order to address the local conditions in the scenario assumption and to ensure that the results can contribute to a low carbon strategy of Wuxi, the quantitative scenario modelling was integrated with a qualitative approach.

In this project three scenarios were developed; however for the presentation of results in this particular article, only two have been selected:

- The Current Policy Scenario (CPS) was developed based on the assumption that;
  - o No additional policies or targets will be adopted beyond the existing policy framework of Wuxi (i.e. a 50% reduction in CO_2 intensity of GDP by 2020 compared to the 2005 level, in accordance with Wuxi Low Carbon City Development Plan);
  - o A reduction in CO_2 intensity of 65%, 70%, and 75% by 2030, 2040 and 2050 respectively, compared to the 2005 level, referring to the baseline scenario in [14].
- The Extra Low Carbon Scenario (ELCS) assumes a significantly more ambitious implementation of low carbon strategies and energy efficient technologies, including "high-hanging fruits" (i.e. those that are more effective but costly).

To create the scenarios a quantitative energy and GHG emission simulation model was used. The modelling approach can be described as a model framework, consisting of one core model and five sub-models. The core model is an energy system simulation model and has been developed by the Wuppertal Institute (WI). The model's database is linked to the sub-models of industry, commerce, households, transport and energy supply. They have been developed as disaggregated technology-based simulation models to account for the use of Low Carbon (LC) technologies. The modelling approach is, therefore, basically bottom-up, without explicit economic optimization. In the CPS, data on industry energy use and economic activity was provided by an econometric projection, whereas the modelling in the ELCS is much more differentiated. Process-sharp modelling of energy-intensive processes allows for the evaluation of individual technologies, which is highly relevant for deriving strategic approaches in the roadmap. The effective production capacity and the age of the manufacturing plants in Wuxi City were evaluated in detail.

The same assumptions about the general socio-economic trends were used for all the scenarios; these assumptions can be characterised by slower yet still significant economic growth and a shift towards more service-oriented products (see Table 1). The specific assumptions were developed based on the China 2050 scenario developed by the Energy Research Institute [14]. In order to adapt the nationwide assumptions to Wuxi's situation, the modeller team consulted local stakeholders and experts about whether certain low carbon technologies were tailored to local conditions and, therefore, applicable in Wuxi.

Table 1. Socio-economic framework

	Unit	2009	2020	2030	2040	2050	2009-2050 [%] growth p.a.
Permanent population	1,000	6,245	6,731	6,598	6,325	5,826	-0.2
GDP	Mill. RMB$_{2005}$	452,175	1,138,622	2,300,540	3,758,442	5,360,192	6.2
Primary sector	Mill. RMB$_{2005}$	8,481	14,217	17,970	21,186	23,775	2.5
Secondary sector	Mill. RMB$_{2005}$	256,933	615,347	1,143,216	1,659,976	2,116,632	5.3
Tertiary sector	Mill. RMB$_{2005}$	186,761	509,057	1,139,355	2,077,280	3,219,784	7.2
Productivity	1,000 RMB$_{2005}$/ work-force	99	230	487	913	1,401	6.7

Source: Wuxi Statistical Yearbook, projections derived by China Environmental Research (CER) from Jiang, et al. (2008) [11, 14]

The assumed quantity of industrial production emphasizes Wuxi's importance as an industrial centre for energy-intensive products. Whereas Wuxi's share in China's population is only 0.5% today, 1.9% of China's crude steel production is concentrated in the Wuxi region. In the scenarios Wuxi does not lost its role as a centre of energy intensive production. The physical production of energy-intensive products such as steel, cement and fertilizers, as well as paper and caustic soda (or chlorine) remains stable (see Table 2). As the overall levels of steel, cement and fertilizer production in China in 2050 are assumed to be lower than today, Wuxi's share of China's production will increase. The production of paper and caustic soda, however, will increase in China, resulting in a lower share for Wuxi.

Table 2. Wuxi's share of energy-intensive sectors in China (2009 and 2050, in physical units and % in the Current Policy Scenario (CPS) (Variant GDP Medium)

	2009	2050	2009	2050	2009	2050
Output [mln t]	China		Wuxi		Share Wuxi [%]	
Steel	572	360	10.6	11.3	1.9	3.1
Cement	1,644	900	15.5	15.5	0.9	1.7
Caustic soda	18	24	0.1	0.1	0.6	0.5
Paper	90	120	0.8	0.8	0.9	0.7
Fertilizer	64	61	0.3	0.3	0.5	0.5
Population [mln]	1,335	1,460	6.2	5.8	0.5	0.4

Source: Wuxi Statistical Yearbook, CER (Wuxi), Jiang, et al. (2008) [11, 14]

Table 3 shows the assumptions about key technologies deployed in the key sectors in the CPS and ELCS. These technologies were identified based on the "Technology Matrix for Germany" developed by the Wuppertal Institute [15, 16] based on literature research and in consultation with local experts from city planning and enterprises in Wuxi taking into account Wuxi's climatic and industrial context.

Table 3. Assumptions for key technologies applied in key sectors in the Current Policy Scenario (CPS) and the Extra Low Carbon Scenario (ELCS)

Key sectors	CPS	ELCS
Industry	Decreasing energy intensity according to global CO_2 intensity targets; very moderate fuel shift	New investments and replacement of old production stock will be equipped with Best Available Technologies (BAT) from 2020 onwards. No substantial energy efficiency improvement will be reached in the short and medium-term in steel and cement industry. Technology improvement of steel production: direct reduction of iron via hydrogen from 2047; Carbon Capture and Storage (CCS) connecting top gas recycling from 2037; smelt reduction process with CCS. Cement industry: cement demand will decline from 2021. Chemical industry: a shift from coal to natural gas in the feedstock; ammonia production will be phased out from 2039.
Power/Heat Generation	Expansion of combined heat and power plants (CHP) as natural gas-fired plants according to industry's thermo power demand. Other new power plants are modelled as conventional coal power plants (ultra-supercritical boilers).	Expansion of combined heat and power plants (CHP) as natural gas-fired plants according to industry's thermo power demand. No further coal-fired power plants will be constructed Increased installation of CHP/CHCP and natural gas power plants. The share of renewable energies will increase (up to 73 GWh in 2050) by: exploiting (limited) local potential: biogas from municipal waste and from agricultural residues, 330 MW wind (installed by 2035), 1,800 MW PV (installed by 2050) and importing renewable energy electricity in addition to local production.
Buildings (residential and commercial)	Heating and cooling appliances are replaced in the old building stock. Refurbishment of the building envelope is not considered. Moderate energy efficiency gains through more efficient household appliances.	Highly energy efficient appliances will be purchased from 2020. Inefficient air conditioners will be replaced. Increasing the share of ultra-low energy and plus energy residential buildings from 2020 (will reach 100% of respective new constructions in 2050).
Transport	Expansion of the subway network according to local plans. Moderate saturation rate of 300 cars per 1000 households is reached in 2025. Rapid growth of freight transport.	Stronger market diffusion of electric vehicles (60% market share of new cars in 2050) compared to CPS. All other assumptions equal to CPS.

Source: Wuppertal Institute based on [15, 16] and expertise of local government and enterprises gained in interviews and online survey.

Table 3 shows from which year on low carbon technologies phase in Wuxi's economy and households in the ELCS. The technology matrix provided information about the availability of technologies on the timeline. Modelling shows necessary new investments in the respective years. So the scope of low carbon technology phase-in can be determined.

Scenario assumptions and results were drafted by the modelling team based on prior work and literature review and discussed with local stakeholders in the context of several expert and stakeholder workshops. The assumptions for key technologies were given to local technology experts who commented on the relevance and applicability of the technologies in Wuxi.

The ELCS can thus be characterized as an explorative scenario rather than a target orientated scenario. It is to show which degree of mitigation can be achieved with technical measures taking into account the socio-economic framework and investment cycles.

GHG INVENTORY RESULTS & KEY SECTORS

GHG inventory results

Of greatest significance for Wuxi are energy-related emissions, which result in the most part from the combustion of raw coal (and only to a limited extend from other fossil fuels, such as diesel). The emissions mainly stem from three different sectors: the energy industry (1A1), the industry sector (1A2) and the transport sector (1A3) (see Figure 1) and are related to the high demand for energy in Wuxi.

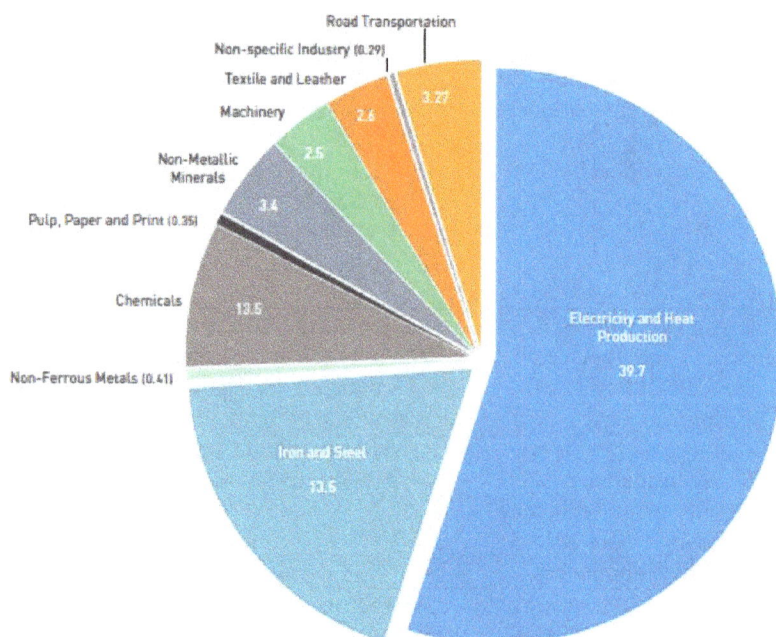

Figure 1. Total CO_2 emissions of Sector 1A, Energy (1A1-1A3), divided into sub-sectors (in mln t CO_2); see also Table 4 for detailed numbers

Electricity and heat production are the main sources of emissions; representing more than half of the current CO_2 emissions at a level of almost 40 million tonnes in 2009 (see Table 4 and Figure 1). In the manufacturing industries, energy-intensive fields such as the iron and steel industry and the chemical industry currently produce a high share of the total CO_2 emissions, but other sectors also produce considerable levels of CO_2.

Wuxi has several industries with potentially high process-related emissions. Although the original intention of the project was to analyse and incorporate all "Kyoto" GHGs, there was insufficient data for these to be calculated comprehensively. Therefore, for non-CO_2 emissions and non-energy related CO_2 emissions, qualitative assumptions or process descriptions have been given. The total emissions as currently reported are higher than the results presented in the inventory [3, 4]. For agriculture, only figures for certain sub-sectors were calculated. Furthermore, it was not possible to produce figures for the waste sector; all of which means that there is the potential for significant improvements to be made in future inventory work.

Table 4. Relevance of proposed key sectors, data availability and improvement potential of calculation methodology, current emissions

IPCC Sectors			Relevance Key sector	Data Availability	EF & Method Improvement potential	CO$_2$ emissions [mln t]	other GHG emissions* [kt CO$_2$e]
1A1		Energy industry	HIGH	++	+	39.7	
	1A1a	Electricity and heat production	HIGH	++	+	39.7	200.0
	1A1b	Petroleum refining	MIDDLE	(+)	++	NE	NE
1A2		Manufacturing industry	HIGH	+	++	29.0	200.28
1A2a/2C1		Iron and steel industry	HIGH	+	++	13.5	91.5
1A2b / 2C3		Non-ferrous metals / Aluminium	HIGH / MIDDLE	+ / --	++	0.41	2.5
1A2c / 2B		Chemical industry	HIGH /	+ / --	++	5.8	40.5
1A2f / 2A1		Non-metallic minerals (cement)	HIGH	(+)	+	3.4	23.6
1A2h/ 2E		Machinery (electronic industry)	MIDDLE	--	+	2.5	17.2
	1A2k	Construction	HIGH	--	++	NE/ND (!)	
	1A2l	Textile and leather	MIDDLE	--	++	2.6	19.4
1A3		Transport	HIGH	(+)	+	3.5	
	1A3b	Road transportation	HIGH	(+)	+	3.27	NE
1A4		Others					
	1A4b	Commercial	HIGH	--	+	IE**	
	1A4b	Residential	HIGH	--	+	IE**	0.58
2		IPPU (other than 1A2)					
	2D	Non-energy from fuels & solvents	MIDDLE	--	++	NE/ND (!)	NE/ND (!)
	2F	Substitute for ozone DS	MIDDLE	--	++	NE/ND (!)	NE/ND (!)
3		AFOLU	MIDDLE	(+)	+		881.2
4		Waste	MIDDLE	--	++	---	NE/ND (!)

Explanation of categories: Relevance: High (key sector); Middle (possible key sector); (++) high potential /sufficient data availability; (+) data available, but not sufficient; (--) no data available; "IE" included elsewhere; "NE" not estimated; "NE/ND (!)" not estimated due to unavailable data/factors; * so far only CH$_4$ & N$_2$O; ** calculated for total consumption for information only; included in electricity production; IPPU = Industrial Processes and Product Use ; AFOLU = Agriculture, Forestry and Other Land use

In 2008 the national government had asked provincial governments to urge their cities to develop emission inventories [17, 18]. And in 2011 the Chinese central government issued "Guidelines on Provincial Greenhouse Gas Emission Inventory (Trial)" which now serves as a basis for developing city inventories [19]. These guidelines are based on both the IPCC Guidelines, which were used in this study and on experiences of inventory development in China in 2005 [20]. Approaches to city specific inventory methodologies, also suitable for emerging and developing countries, have recently been developed (e.g. by WRI) [21]. Moreover, the World Resources Institute, together with several other organizations, developed the GHG Accounting Tool for Chinese Cities (pilot 1.0) to explore methods of measuring GHG emissions in Chinese cities. The tool takes the administrative boundary of Chinese cities as the accounting boundary and combines a top-down and bottom-up data collection approach to address the significant data gap [22]. Wuxi is intending to have a regular monitoring system, which is currently being developed.

Key sectors identified

The most relevant key sectors from the perspective of current GHG emissions were identified, taking into account their relevance for regional policies, vulnerability to climate change and (past and) future trends. The relevance of the selected emission sectors, their current CO_2, CH_4 and N_2O emissions, as well as the quality of available data and calculation methodology, are presented in Table 4.

In addition to the dominant power and heat sectors, most of the key sectors are industry sectors such as the iron and steel industry, chemical industry, non-metallic minerals (cement), as well as electrical equipment and machinery manufacturing. In addition, the construction sector is relevant due to its high material/resource use and the increasing demand for electricity for cooling and heating purposes (especially where the design and construction of buildings inhibits the efficient use of energy flows). The latter is also relevant to adaptation needs due to an increase in both temperatures and living standards. Linked to this is the ever-growing requirement for electricity in the residential sector and in the commercial (service) sector, which are also regarded as two relevant key sectors. Road transport is as well considered as key sector due to the remarkable increase in number of vehicles, its related emissions and air pollution, and the future infrastructural challenges.

SCENARIO RESULTS

Energy consumption and total emissions in the Current Policy Scenario (CPS) and the Extra Low Carbon Scenario (ELCS)

Figure 2 illustrates the development of Wuxi's primary energy consumption by energy source in the two scenarios. The most striking feature of the ELCS in comparison to the CPS is the 90% reduction in coal consumption compared to today's level, an 85% reduction is achieved solely in the years after 2040. The most prominent reason is that the existing coal power plants gradually cease production and are not replaced by new coal in the ELCS. The remaining coal use in the industry sector in the ELCS in 2050 is partly due to the production of steel with coal in combination with carbon capture and storage (CCS). The large proportion of natural gas is due to its role as a bridge to renewable electricity production and the local demand of thermo power which cannot be provided by limited biomass resources.

Figure 3 compares the final energy demand in the two scenarios. Thereby the demand for energy of all sectors is illustrated, i.e. direct demand for fossil fuels (e.g. in the heavy industry or transport sector) and the demand for thermo power, hydrogen and electricity. According to the definitions of energy statistics derived energy carriers like hydrogen, thermo power and electricity are not classified as renewable here, although they can be produced from renewable sources. The actual share of renewables can be seen in Figure 2. Final energy demand is projected to increase consistently in both scenarios until 2020. Only from 2030 onwards does the ELCS show a reduction of final energy demand. The ever-growing share of electricity is caused mainly by the rapid growth of the service sector and new industries in Wuxi, such as machinery construction, which consume less heat but are heavily reliant on electricity. Heat demand in industry is, to a large degree, provided by Combined Heat and Power plants (CHP) that are currently fired by coal, but will be replaced in the scenarios by gas-fired units. The use of hydrogen, as a new energy carrier in 2050, is limited to steel production in the ELCS, where it serves as a reducing agent in the direct reduction of iron ore.

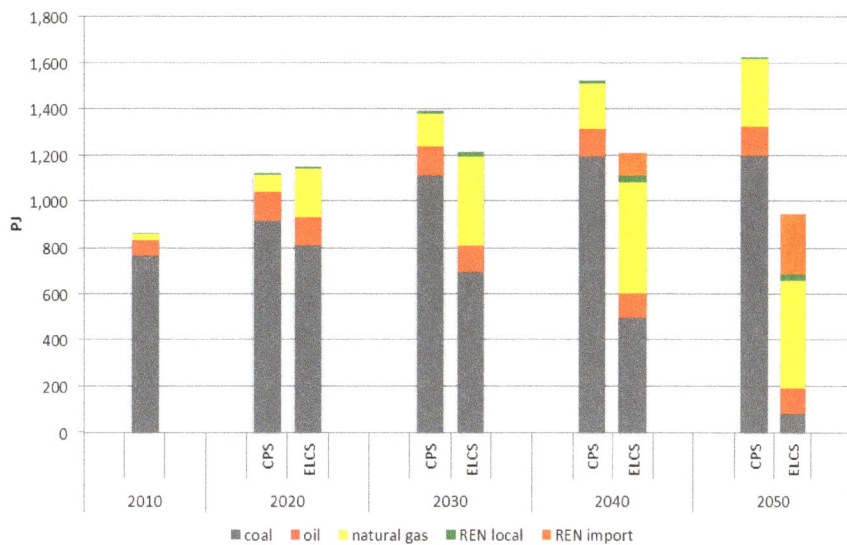

Figure 2. Comparison of primary energy consumption in the Current Policy Scenario (CPS) and the Extra Low Carbon Scenario (ELCS) (2010-2050)

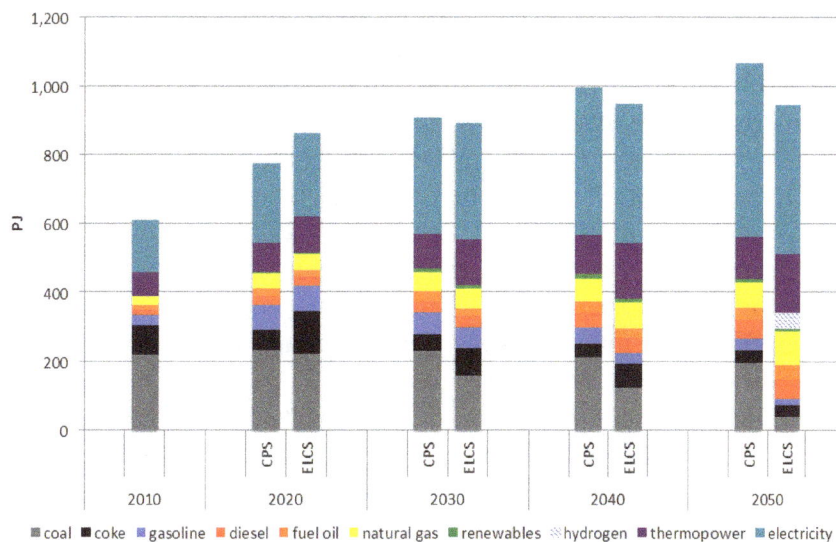

Figure 3. Comparison of final energy use in the Current Policy Scenario (CPS) and the Extra Low Carbon Scenario (ELCS) (2010-2050)

Figure 4 shows CO_2 emission trends (including direct and indirect emissions) in the two scenarios start to deviate considerably after 2020, because the general assumption before 2020 was based on the existing policy framework adopted by Wuxi government. CO_2 emissions in the ELCS will be significantly lower than in the CPS, due to the employment of Best Available Technologies (BAT), a fuel shift towards gas and renewable electricity as well as highly effective mitigation measures in all key sectors. In 2050, CO_2 emissions in the ELCS will be 36 million tonnes, equivalent to a quarter of that in the CPS (140 million tonnes) and a reduction of 56% compared to 2010.

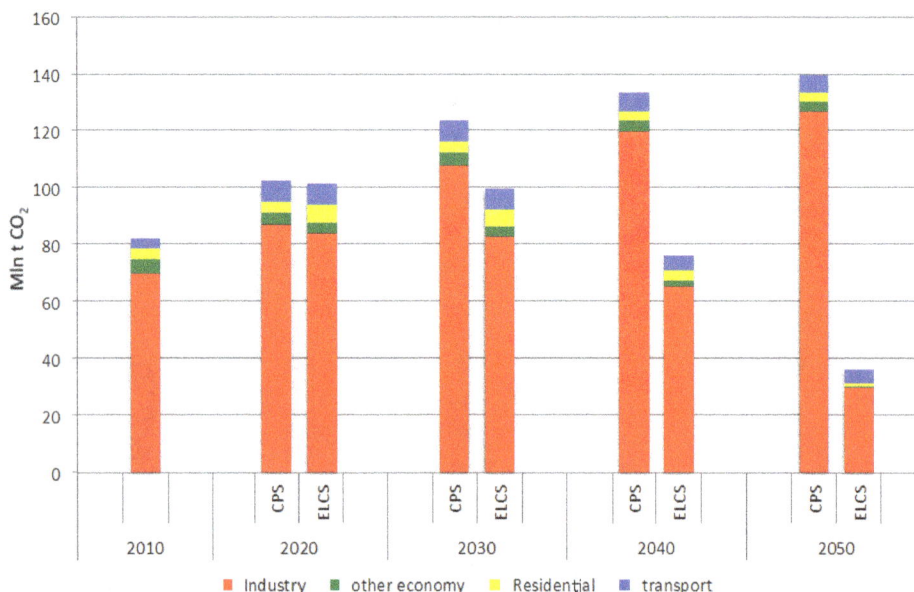

Figure 4. Comparison of direct and indirect CO_2 emissions of sectors (including indirect CO_2 emissions allocated to thermo power, heat and electricity) in the Current Policy Scenario (CPS) and the Extra Low Carbon Scenario (ELCS) (2010-2050)

CONCLUSIONS AND OUTLOOK

In the Low Carbon Future Cities study, important steps were taken to gain reliable information on the current situation and possible future development in Wuxi with regard to its sustainability. The picture gained from the the GHG inventory and scenario analysis is impressive and leads to the conclusion that enormous efforts, on different levels, will be necessary to drive Wuxi towards a low carbon path in the future.

Looking at the per capita emissions in Wuxi, current levels are already high at around 12 tonnes CO_2 per capita (2009, permanent population according to census data) compared to Western European cities (such as Düsseldorf or London) at less than 6 tonnes CO_2 per capita [3]. Although Wuxi has set reduction targets and developed a low carbon plan, the projected results based on the current policies (CPS) show that without more ambitious efforts, the total emissions would increase to 23.6 tonnes CO_2 per capita by 2050. However, if the ELCS pathway was to be adopted, these CO_2 emission levels could be reduced to 6.4 tonnes per capita by 2050 (half of the current levels). This level is, however, still very high despite the already supportive current policies and additional significant technology improvements in the ELCS scenario, even taking into account "high-hanging fruit" technologies (costly and yet to be developed). However, it is worth noting that Wuxi has, and is expected to continue to have, an important role in China's economy, producing a number of energy-intensive goods for the regional, national and international markets.

If we take the ELCS assumptions for Wuxi as a benchmark for China, China will reach an emission level of 2.9 tonnes per capita, which is still above the level necessary to comply with the 2 degree target, following expert analyses. The German Advisory Council on Global Change [23] asserts that industrial and emerging countries should achieve about one tonne of CO_2 per capita by 2050. To reduce CO_2 emissions further, other mitigation measures should be considered; for example, modal shift for freight transport that was not considered in the scenario. Further CO_2 emission reductions will require an overall reduction in resource use, such as steel and cement.

Despite some limitations of the inventory and scenario results presented due to data restrictions, lack of detailed data in certain sectors and uncertainties in the scenario assumptions, the results can be regarded as a good basis for the development of a roadmap towards a low carbon city strategy for Wuxi.

In the analysis of the current and future GHG emissions of Wuxi it became obvious that such information is important to understand the impacts of energy and resource use in different sectors, as well as the underlying and future drivers. However, these types of stand-alone studies and analyses have limitations, are time-consuming and can only be regarded as the first steps. Regular monitoring of current performance is crucial to measure the impacts of the implemented measures and to identify possible new hot spots.

As urban growth is expected to be the highest in "medium-sized" cities and not in the metropolises, which are frequently analysed [15], this study can also serve as an example and basis for discussion for other industrial cities in China of similar size and with similar conditions. As stated in this paper, the implementation of comparative tools at city level and regular accounting are also imperative. In addition to the territorially produced GHG emissions, which are often in the focus of global political discussions, other aspects and impacts are also crucial and need to be considered when aiming for sustainability. This includes not only local activities, but also the external impact of resources used by the city of Wuxi and the city's vulnerability to climate change. Both these dimensions are considered in the LCFC study and are detailed in the specific background papers [3-5].

Overall, it is clear from the modelling that it would be difficult to initiate and implement a carbon neutral strategy at local level in isolation. Wuxi is an industrial Chinese city that is export oriented and answering the global demand for products. The dominant energy and emission source in China is coal and one city alone cannot easily reduce or even step out of its use. However, this is not a problem unique to Wuxi or China. A comprehensive rethink at global level on how to increase efficiency and sustainability is required. China, and its cities such as Wuxi, have an important role to play and can generate a remarkable impact; however, longstanding industrial countries in Europe and North America need to take the leading role in this global process.

ACKNOWLEDGEMENTS

The project consortium consists of five Chinese and two German partner organizations and cooperates closely with the city governments of the two regions Wuxi and Düsseldorf. The study is supported by the German organization Stiftung Mercator.

The authors would like to thank Andreas Oberheitmann from China Environment Research (CER), as well as Wang Haoping and Wang Can from the Research Centre for International Environmental Policy (RCIEP) at Tsinghua University, for their cooperation and contributions to the project. Special thanks goes to our colleague Mathieu Saurat, who was responsible for the material use models, and all our colleagues from the Wuppertal Institute who have been involved with the project. We also would like to thank our other colleagues in the consortium for their productive cooperation and helpful information: Sui Yingying and others from the Wuxi Low Carbon Center

(WLCC), Thomas Fischer, Marco Gemmer and Jiang Tong from China Meteorological Administration (CMA), National Climate Centre (NCC), Neil Cole and Sebastian Philipps from the Collaborating Centre on Sustainable Consumption and Production (CSCP) and our colleagues from the Chinese Society for Sustainable Development, Low Carbon Group (CSSD). In addition, we would like to thank the city of Wuxi for their good cooperation in this project.

We would also like to acknowledge the financial support provided to the LCFC project by Siftung Mercator and thank them for making this study possible.

The paper is based on background information in the Work Packages 2, 3 and 5 [3, 5, 6], where details on the results can be found.

REFERENCES

1. United Nations Human Settlements Programme, *Cities and Climate Change (Global Report on Human Settlements 2011)*, UN-Habitat, 2011, xviii + 279 pp, London, Earthscan/Routledge, 2011.

2. Vallentin, D., et al., Specification of the Pilot Studies and Methodological Framework Scoping Paper - WP1 Low Carbon Future Cities Project Wuppertal, 2011, http://www.lowcarbonfuture.net. [Accessed: 17- Semptember -2014]

3. Dienst, C., et al., WP2 Report - Integrated Status Quo and Trends assessment in Wuxi, Overview of WP2 results Low Carbon Future Cities Report, Wuppertal, 2012, http://www.lowcarbonfuture.net. [Accessed: 17- Semptember -2014]

4. Dienst, C., Schneider, C., Xia, C., Saurat, M., Fischer, T., Vallentin, D., On Track to become a Low Carbon Future City? First Findings from the Pilot City of Wuxi, *Sustainability*, Vol. 5, No. 8, pp 3224-3243, 2013.

5. Venjakob, J. and Schneider, C., Integrated City Strategy for CO_2-Emission Reduction, Resource Efficiency and Climate Resilience, *Final Report of Work Package 3*, Low Carbon Future Cities Report, Wuppertal, 2013.

6. Vallentin, D., Dienst, C., Xia-Bauer, C., From Scenarios to Action - Facilitating a Low Carbon Pathway for Wuxi, Needs / Possible solutions / Measures, Wuppertal, 2013.

7. Yetano-Roche, M., Lechtenböhmer, S., Fischedick, M., Gröne, M-C., Xia, C., Dienst, C., Concepts and Methodologies for Measuring the Sustainability of Cities, Annual Review of Environment and Resources, Vol. 39, (in press, Volume publication date November 2014).

8. Wang, H., et al., Mitigating Greenhouse Gas Emissions from China's Cities: Case Study of Suzhou, *Energy Policy*, Vol. 68, pp 482-489.

9. Cai, B., Zhang, L., Urban CO_2 emissions in China: Spatial Boundary and Performance Comparison, *Energy Policy*, Vol. 66, pp 557-567, 2014.

10. IPCC (Intergovernmental Panel on Climate Change). IPCC Guidelines for National Greenhouse Gas Inventories, Volume 1-5, Japan: IGES, 2006.

11. WMBS (Wuxi Municipal Bureau of Statistics, 2001-2010), Wuxi Yearbook 2001 - Wuxi Yearbook 2010, Beijing: China Statistics Press, http://www.wxtj.gov.cn/tjxx/tjsj/tjnj//index.shtml. [Accessed: 28-March-2011]

12. Marland, G., Emissions Accounting: China's Uncertain CO_2 emissions, *Nature Climate Change*, Vol. 2, No. 11, pp 645-646, 2012.

13. Reuters, China emissions Study suggests Climate Change could be faster than thought, http://www.reuters.com/article/2012/06/10/us-china-emissions-idUSBRE8590AD2012 0610. [Accessed: 17-Semptember-2014]

14. Jiang, K. J., Hu, X. L., Zhuang, X., et al., China's Energy Demand and Greenhouse Gas Emission Scenarios in 2050, *Advances in Climate Change Research*, Vol. 4, No. 5, pp 296-302, 2008.

15. Landeshauptstadt Düsseldorf (ed.), Technologiematrix Deutschland, *Technologieoptionen für klimaverträgliche Großstädte 2050*, Report funded by the German Federal Ministry of Environment and the City of Düsseldorf, Düsseldorf, January 2011.

16. Lechtenböhmer, S., et al., Redesigning Urban Infrastructures for a Low Emission Future, A Technology Overview, *Surveys and Perspectives Integrating Environment and Society*, Vol. 3, No. 2, pp 1-16, 2010.

17. Cai, B., Research on Greenhouse Gas Emissions Inventory in the Cities of China, *China Population Resources and Environment*, Vol. 22, pp 21-27, 2012.

18. Cheng, H.-H., Shen, J.-F., Tan, C.-S., CO_2 Capture from Hot Stove Gas in Steel making Process, *International Journal of Greenhouse Gas Control*, Vol. 4, pp 525-531, 2010.

19. NDRC, *China's Policies and Actions for addressing Climate Change*, The Progress Report, 2011.

20. Bai, H., Zeng, S., Dong, X., Chen, J., Substance Flow analysis for an Urban drainage System of a Representative Hypothetical City in China, *Front. Environ. Sci. Eng.*, Vol. 7, pp 746-755, 2013.

21. WRI (World Resource Institute), For the First Time, a Common Framework for Cities' Greenhouse Gas Inventories, http://insights.wri.org/news/2013/03/first-time-common-framework-cities-greenhouse-gas-inventories. [Accessed: 14-May-2013]

22. WRI/WBSCD, Greenhouse Gas Protocol, http://www.ghgprotocol.org/. [Accessed: 01-Semptember-2014]

23. WBGU, Solving the Climate Dilemma: The Budget Approach, Special Report 2009, 58 p. WBGU, German Advisory Council on Global Change, Berlin, http://www.wbgu.de/fileadmin/templates/dateien/veroeffentlichungen/sondergutachten/sn2009/wbgu_sn2009_en.pdf.

Permissions

All chapters in this book were first published in JSDEWES, by SDEWES Centre; hereby published with permission under the Creative Commons Attribution License or equivalent. Every chapter published in this book has been scrutinized by our experts. Their significance has been extensively debated. The topics covered herein carry significant findings which will fuel the growth of the discipline. They may even be implemented as practical applications or may be referred to as a beginning point for another development.

The contributors of this book come from diverse backgrounds, making this book a truly international effort. This book will bring forth new frontiers with its revolutionizing research information and detailed analysis of the nascent developments around the world.

We would like to thank all the contributing authors for lending their expertise to make the book truly unique. They have played a crucial role in the development of this book. Without their invaluable contributions this book wouldn't have been possible. They have made vital efforts to compile up to date information on the varied aspects of this subject to make this book a valuable addition to the collection of many professionals and students.

This book was conceptualized with the vision of imparting up-to-date information and advanced data in this field. To ensure the same, a matchless editorial board was set up. Every individual on the board went through rigorous rounds of assessment to prove their worth. After which they invested a large part of their time researching and compiling the most relevant data for our readers.

The editorial board has been involved in producing this book since its inception. They have spent rigorous hours researching and exploring the diverse topics which have resulted in the successful publishing of this book. They have passed on their knowledge of decades through this book. To expedite this challenging task, the publisher supported the team at every step. A small team of assistant editors was also appointed to further simplify the editing procedure and attain best results for the readers.

Apart from the editorial board, the designing team has also invested a significant amount of their time in understanding the subject and creating the most relevant covers. They scrutinized every image to scout for the most suitable representation of the subject and create an appropriate cover for the book.

The publishing team has been an ardent support to the editorial, designing and production team. Their endless efforts to recruit the best for this project, has resulted in the accomplishment of this book. They are a veteran in the field of academics and their pool of knowledge is as vast as their experience in printing. Their expertise and guidance has proved useful at every step. Their uncompromising quality standards have made this book an exceptional effort. Their encouragement from time to time has been an inspiration for everyone.

The publisher and the editorial board hope that this book will prove to be a valuable piece of knowledge for researchers, students, practitioners and scholars across the globe.

List of Contributors

Alejandro del Amo Sancho
Mechanical Department Universidad de Zaragoza, Zaragoza, Spain

Fernando Amaral de Almeida Prado Jr.
Faculdade de Engenharia Civil UNICAMP, Sao Paulo, Brazil

Ana Lúcia Rodrigues da Silva
Senac Campos do Jordão, Sao Paulo, Brazil

Edvaldo Marcelo Avila
Comerc Energia, Sao Paulo, Brazil

Gustavo Matsuyama
Sinerconsult Consultoria Treinamento e Participações Limitada, Sao Paulo, Brasil

Jorge Morel
Department of Electrical and Electronic Engineering, Kitami Institute of Technology, Kitami, Japan

Shin'ya Obara
Department of Electrical and Electronic Engineering, Kitami Institute of Technology, Kitami, Japan

Yuta Morizane
Department of Electrical and Electronic Engineering, Kitami Institute of Technology, Kitami, Japan

Bin-Juine Huang
Department of Mechanical Engineering, National Taiwan University, Taipei, Taiwan

Tze-Ling Chong
1Department of Mechanical Engineering, National Taiwan University, Taipei, Taiwan

Hsien-Shun Chang
Department of Mechanical Engineering, National Taiwan University, Taipei, Taiwan

Po-Hsien Wu
Department of Mechanical Engineering, National Taiwan University, Taipei, Taiwan

Yeong-Chuan Kao
Department of Physics, National Taiwan University, Taipei, Taiwan

Hajime Sasaki
Policy Alternatives Research Institute, The University of Tokyo, 7-3-1, Hongo, Bunkyo-ku, Tokyo, Japan

Ichiro Sakata
Policy Alternatives Research Institute, The University of Tokyo, 7-3-1, Hongo, Bunkyo-ku, Tokyo, Japan

Weerin Wangjiraniran
Energy Research Institute, Chulalongkorn University, 12th Floor Institute Building III, Phyathai Road, Pratumwan, Bangkok, Thailand

Sengprasong Phrakonkham
Faculty of Engineering, National University of Laos, Don Noun, Vientiane, Lao PDR

Alessandro Chiodi
Environment Research Institute, University College Cork, Cork, Ireland

Paul Deane
Environment Research Institute, University College Cork, Cork, Ireland

Maurizio Gargiulo
E4SMA, S.r.l., Torino, Italy

Brian Ó Gallachóir
Environment Research Institute, University College Cork, Cork, Ireland

Huili Zhang
Department of Chemical Engineering, Chemical and Biochemical Process Technology and Control Section, KU Leuven, Heverlee, Belgium

Jan Baeyens
College of Life Science and Technology, Beijing University of Chemical Technology, Beijing, China

Jan Degreve
KU Leuven, Heverlee, Belgium

Agnes Gerse
Department of Energy Engineering, Budapest University of Technology and Economics, MAVIR ZRt., Budapest, Hungary

Jorge Morel
Department of Electrical and Electronic Engineering, Kitami Institute of Technology, Japan

Shinýa Obara
Department of Electrical and Electronic Engineering, Kitami Institute of Technology, Japan

Yuta Morizane
Department of Electrical and Electronic Engineering, Kitami Institute of Technology, Japan

Susana Boeykens
Heterogeneous Systems Chemistry Laboratory, Faculty of Engineering, University of Buenos Aires, P. Colon 850 BA, Buenos Aires, Argentina

C. Alejandro Falcó
Heterogeneous Systems Chemistry Laboratory, Faculty of Engineering, University of Buenos Aires, P. Colon 850 BA Buenos Aires, Argentina

Maria Macarena Ruiz Vázquez
Heterogeneous Systems Chemistry Laboratory, Faculty of Engineering, University of Buenos Aires, P. Colon 850 BA, Buenos Aires, Argentina

María Del Carmen Tortorelli
Ecotoxicology Research Program, Department of Basic Sciences, National University of Luján, Ruta 5 y Avenida Constitución, Buenos Aires, Argentina

Giuliano Buceti
UTFUS, ENEA, C.R. Frascati, C.P. 65, I-00044, Frascati, Italy

Farajallah Alrashed
School of Engineering and Built Environment, Glasgow Caledonian University, Glasgow, UK

Muhammad Asif
Department of Architectural Engineering, King Fahd University of Petroleum & Minerals, Dhahran, Saudi Arabia

Tomas Losak
Department of Environmentalistics and Natural Resources, Faculty of Regional Development and International Studies, Mendel University in Brno, Brno, Czech Republic

Jaroslav Hlusek
Department of Environmentalistics and Natural Resources, Faculty of Regional Development and International Studies, Mendel University in Brno, Brno, Czech Republic

Andrea Zatloukalova
Department of Agrochemistry, Soil Science, Microbiology and Plant Nutrition, Faculty of Agronomy, Mendel University in Brno, Brno, Czech Republic

Ludmila Musilova
Department of Agrochemistry, Soil Science, Microbiology and Plant Nutrition, Faculty of Agronomy, Mendel University in Brno, Brno, Czech Republic

Monika Vitezova
Department of Agrochemistry, Soil Science, Microbiology and Plant Nutrition, Faculty of Agronomy, Mendel University in Brno, Brno, Czech Republic

Petr Skarpa
Department of Agrochemistry, Soil Science, Microbiology and Plant Nutrition, Faculty of Agronomy, Mendel University in Brno, Brno, Czech Republic

Tereza Zlamalova
Department of Agrochemistry, Soil Science, Microbiology and Plant Nutrition, Faculty of Agronomy, Mendel University in Brno, Brno, Czech Republic

Jiri Fryc
Department of Agricultural, Food and Environmental Engineering, Faculty of Agronomy, Mendel University in Brno, Brno, Czech Republic

Tomas Vitez
Department of Agricultural, Food and Environmental Engineering, Faculty of Agronomy, Mendel University in Brno, Brno, Czech Republic

Jan Marecek
Department of Agricultural, Food and Environmental Engineering, Faculty of Agronomy, Mendel University in Brno, Brno, Czech Republic

Anna Martensson
Department of Soil and Environment, Swedish University of Agricultural Sciences, Uppsala, Sweden

Sadik Bekteshi
Faculty of Natural Sciences, University of Prishtina "Hasan Prishtina", Prishtina, Kosovo

Skender Kabashi
Faculty of Natural Sciences, University of Prishtina "Hasan Prishtina", Prishtina, Kosovo

Skender Ahmetaj
Faculty of Natural Sciences, University of Prishtina "Hasan Prishtina", Prishtina, Kosovo

Ivo Šlaus
Ruđer Bošković Institute, Bijenička 54, Zagreb, Croatia

Aleksander Zidanšek
Jožef Stefan International Postgraduate School, Faculty of Natural Sciences and Mathematics, University of Maribor, Jamova 39, Ljubljana, Slovenia

Kushtrim Podrimqaku
Faculty of Electrical and Computer Engineering, University of Prishtina, "Hasan Prishtina", Mother Teresa Str., Prishtina, Kosovo

Shkurta Kastrati
Department of Power Engineering Equipment, Technická univerzita, Studentská 1402/2, Liberec, Czech Republic

Sebastian Goers
Energy Institute at the Johannes Kepler University, Linz, Austria, Department of Energy Economics

Carmen Dienst
Wuppertal Institute for Climate, Environment and Energy GmbH, Wuppertal, Germany

Chun Xia
Wuppertal Institute for Climate, Environment and Energy GmbH, Wuppertal, Germany

Clemens Schneider
Wuppertal Institute for Climate, Environment and Energy GmbH, Wuppertal, Germany

Daniel Vallentin
Wuppertal Institute for Climate, Environment and Energy GmbH, Wuppertal, Germany

Johannes Venjakob
Wuppertal Institute for Climate, Environment and Energy GmbH, Wuppertal, Germany

Ren Hongyan
Jiangnan University, Environment and Civil engineering school, Wuxi Low Carbon Development Research Centre (WLCDC), Wuxi, China